Transforming the Internet of Things for Next-Generation Smart Systems

Bhavya Alankar
Jamia Hamdard, India

Harleen Kaur
Hamdard University, India

Ritu Chauhan
Amity University, India

A volume in the Advances in Computational Intelligence and Robotics (ACIR) Book Series

Published in the United States of America by
IGI Global
Engineering Science Reference (an imprint of IGI Global)
701 E. Chocolate Avenue
Hershey PA, USA 17033
Tel: 717-533-8845
Fax: 717-533-8661
E-mail: cust@igi-global.com
Web site: http://www.igi-global.com

Library of Congress Cataloging-in-Publication Data

Names: Alankar, Bhavya, 1981- editor. | Kaur, Harleen, editor. | Chauhan, Ritu, 1983- editor.
Title: Transforming the internet of things for next-generation smart systems / Bhavya Alankar, Harleen Kaur, and Ritu Chauhan, editors.
Description: Hershey, PA : Engineering Science Reference, an imprint of IGI Global, [2021] | Includes bibliographical references and index. | Summary: "With the Internet-of-Things (IoT) showing massive potential to transform current business models and enhance human lifestyles, this book investigates the abundance of knowledge being generated through entirely new eco-systems of information generating applications in various realms such as pervasive healthcare, smart homes, smart cities, connected logistics, automated supply-chain, manufacturing units, and many more areas"-- Provided by publisher.
Identifiers: LCCN 2020054410 (print) | LCCN 2020054411 (ebook) | ISBN 9781799875413 (hardcover) | ISBN 9781799875420 (paperback) | ISBN 9781799875437 (ebook)
Subjects: LCSH: Internet of things. | Automatic control. | Artificial intelligence.
Classification: LCC TK5105.8857 .T74 2021 (print) | LCC TK5105.8857 (ebook) | DDC 004.67/8--dc23
LC record available at https://lccn.loc.gov/2020054410
LC ebook record available at https://lccn.loc.gov/2020054411

This book is published in the IGI Global book series Advances in Computational Intelligence and Robotics (ACIR) (ISSN: 2327-0411; eISSN: 2327-042X)

Advances in Computational Intelligence and Robotics (ACIR) Book Series

Ivan Giannoccaro
University of Salento, Italy

ISSN:2327-0411
EISSN:2327-042X

MISSION

While intelligence is traditionally a term applied to humans and human cognition, technology has progressed in such a way to allow for the development of intelligent systems able to simulate many human traits. With this new era of simulated and artificial intelligence, much research is needed in order to continue to advance the field and also to evaluate the ethical and societal concerns of the existence of artificial life and machine learning.

The **Advances in Computational Intelligence and Robotics (ACIR) Book Series** encourages scholarly discourse on all topics pertaining to evolutionary computing, artificial life, computational intelligence, machine learning, and robotics. ACIR presents the latest research being conducted on diverse topics in intelligence technologies with the goal of advancing knowledge and applications in this rapidly evolving field.

COVERAGE

- Agent technologies
- Evolutionary Computing
- Adaptive and Complex Systems
- Computational Intelligence
- Cognitive Informatics
- Computational Logic
- Artificial Intelligence
- Brain Simulation
- Artificial Life
- Natural Language Processing

IGI Global is currently accepting manuscripts for publication within this series. To submit a proposal for a volume in this series, please contact our Acquisition Editors at Acquisitions@igi-global.com or visit: http://www.igi-global.com/publish/.

Titles in this Series

For a list of additional titles in this series, please visit: www.igi-global.com/book-series

Machine Learning Techniques for Pattern Recognition and Information Security
Mohit Dua (National Institute of Technology Kurukshetra, India) and Ankit Kumar Jain (National Institute of Technology, Kurukshetra, India)
Engineering Science Reference • © 2021 • 300pp • H/C (ISBN: 9781799832997) • US $225.00

Driving Innovation and Productivity Through Sustainable Automation
Ardavan Amini (EsseSystems, UK) Stephen Bushell (Bushell Investment Group, UK) and Arshad Mahmood (Birmingham City University, UK)
Engineering Science Reference • © 2021 • 275pp • H/C (ISBN: 9781799858799) • US $245.00

Examining Optoelectronics in Machine Vision and Applications in Industry 4.0
Oleg Sergiyenko (Autonomous University of Baja California, Mexico) Julio C. Rodriguez-Quiñonez (Autonomous University of Baja California, Mexico) and Wendy Flores-Fuentes (Autonomous University of Baja California, Mexico)
Engineering Science Reference • © 2021 • 346pp • H/C (ISBN: 9781799865223) • US $215.00

Emerging Capabilities and Applications of Artificial Higher Order Neural Networks
Ming Zhang (Christopher Newport University, USA)
Engineering Science Reference • © 2021 • 346pp • H/C (ISBN: 9781799835639) • US $225.00

Machine Learning Applications in Non-Conventional Machining Processes
Goutam Kumar Bose (Haldia Institute of Technology, India) and Pritam Pain (Haldia Institute of Technology, India)
Engineering Science Reference • © 2021 • 313pp • H/C (ISBN: 9781799836247) • US $195.00

Artificial Neural Network Applications in Business and Engineering
Quang Hung Do (University of Transport Technology, Vietnam)
Engineering Science Reference • © 2021 • 275pp • H/C (ISBN: 9781799832386) • US $245.00

Multimedia and Sensory Input for Augmented, Mixed, and Virtual Reality
Amit Kumar Tyagi (Research Division of Advanced Data Science, Vellore Institute of Technolgy, Chennai, India)
Engineering Science Reference • © 2021 • 310pp • H/C (ISBN: 9781799847038) • US $225.00

701 East Chocolate Avenue, Hershey, PA 17033, USA
Tel: 717-533-8845 x100 • Fax: 717-533-8661
E-Mail: cust@igi-global.com • www.igi-global.com

Editorial Advisory Board

Table of Contents

Detailed Table of Contents

Chapter 1

    Nalina Suresh, University of Namibia, Namibia
    Valerianus Hashiyana, University of Namibia, Namibia
    Martin Mabeifam Ujakpa, International University of Management, Namibia
    Anton Limbo, University of Namibia, Namibia
    Gloria E. Iyawa, Namibia University of Science and Technology, Namibia
    Ntinda Maria Ndapewa, University of Namibia, Namibia

The term "cloud of things" is currently in the forefront of computer research topics due to its vital role towards the internet of things. To integrate cloud computing and internet of things into a single technology or worldview, this chapter discussed the evolution of internet of things and cloud computing and reviewed literature on cloud computing and internet of things and their possible integration. The chapter also discussed the importance of cloud internet of things, its architecture, and operation; the need to integrate internet of things and cloud computing; and challenges of cloud internet of things. The chapter then used the identified open issues and future direction to propose a generic cloud internet of things architecture and pilot simulated the results to verify the possibility and effectiveness of cloud computing and internet of things (CIoTs) integration. The researchers believe that the chapter will provide a better insight for anyone who wishes to carry out research in the field of cloud internet of things.

Chapter 2

    Rachna Jain, JSS Academy of Technical Education, Noida, India

Internet of things (IoT) networks is the buzzword these days in Industry 4.0. IoT nodes are resource constrained and should be light enough to minimize the power consumption. IoT paradigm does not depend on human intervention at each and every step. There is a need of "trust" between communicating entities. Devices at physical layer are vulnerable to various attacks such as denial of service (DoS) attack, wormhole attack, etc. Trust becomes more important when vulnerability of attacks increases to the devices. This establishment of trust helps in handling risks in a controlled way in unpredicted situations as well as providing better services at infrastructure level. Social environments can evaluate trust while seeing the relationship between interacting parties; however, in service-oriented industries quality of service (QoS) parameters must be maintained while evaluating trust. So, in this chapter a unique metric expected transmission count (ETX) is employed for implementing QoS while evaluating trust between interacting entities using Cooja simulator.

Internet of things (IoT) guarantees an incredible future for the internet where the sort of correspondence is machine-machine (M2M). This arising standard of networking will impact all aspects of lives going from the computerized houses to smart IoT-based systems by implanting knowledge into the articles. This chapter intends to give an exhaustive outline of the IoT, IIoT situation and audits its empowering innovations. And finally, applications resulting from IoT/IIoT that facilitate daily needs are discussed.

Energy consumption has become a prime concern in designing wireless sensor networks (WSN) for the internet of things (IoT) applications. Smart cities worldwide are executing exercises to progress greener and safer urban situations with cleaner air and water, better adaptability, and capable open organizations. These exercises are maintained by progresses like IoT and colossal information examination that structure the base for smart city model. The energy required for successfully transmitting a packet from one node to another must be optimized so that the average energy gets reduced for successful transmission over a channel. This chapter has been devised to optimize the energy required for transmitting a packet successfully between two communicating sensor nodes using particle swarm optimization (PSO). In this chapter, the average energy for successfully transmitting a packet from one node to another has been optimized to achieve the optimal energy value for efficient communication over a channel. The power received by the sensor node has also been optimized.

The world is going through growth in smart cities, and this is possible because of a revolution of information technology contributing towards social and economic changes and hence endowing challenges of security and privacy. At present, everything is connected through internet of things in homes, transport, public progress, social systems, etc. Nevertheless, they are imparting incomparable development in standard of living. Unified structure commits to welfare, well-being, and protection of people. This chapter surveys two consequential threats, that is, privacy and security. This chapter puts forward review of some paperwork done before consequently finding the contributions made by author and what subsequent work can be carried out in the future. The major emphasis is on privacy security of smart cities and how to overcome the challenges in achievement of protected smart city structure.

Chapter 6

Transforming IT systems educational applications has become imperative in a rapidly evolving global scenario. Today, educational organizations have to provide transparent, confident, secured information and quality data for monitoring and advanced predictive capabilities services to society. Educational organizations have to meet these objectives consistently during typical and crisis scenarios. Modern educational applications are integrated with social network sites, sensors, intelligent devices, and cloud platforms. Hence, data management solutions serve as the basis for educational organizations' information needs. However, modern technologies demand a re-engineer of these platforms to meet the ever-growing demand for better performance, scalability, and educational organizations' availability. This chapter discusses the challenges inherent to the existing educational data system, the architectural methods available to address the above challenges, and the roadmap for building next-generation educational data ecosystems.

Chapter 7

The IoT or the internet of things started as a technology to connect everyday objects over the internet, which has evolved into something big and invaded into every single aspect of our lives. As technology is gaining momentum, IoT-based smart devices usage among users is expanding, which generates massive data at our disposal across various domains. The authors have systematically studied the taxonomy of data analytics and the benefits of using advanced machine learning techniques in converting data into valuable assets. In the studies, they have identified and did due diligence on different smart home systems, their features, and configuration. During this course of study, they have also identified the vulnerability of such a system and threats associated with these vulnerabilities in a secure smart home environment.

    Anton Limbo, University of Namibia, Namibia
    Nalina Suresh, University of Namibia, Namibia
    Set-Sakeus Ndakolute, University of Namibia, Namibia
    Valerianus Hashiyana, University of Namibia, Namibia
    Titus Haiduwa, University of Namibia, Namibia
    Martin Mabeifam Ujakpa, International University of Management, Namibia

Farmers in Namibia currently operate their irrigation systems manually, and this seems to increase labor and regular attention, especially for large farms. With technological advancements, the use of automated irrigation could allow farmers to manage irrigation based on a certain crops' water requirements. This chapter looks at the design and development of a smart irrigation system using IoT. The conceptual design of the system contains monitoring stations placed across the field, equipped with soil moisture sensors and water pumps to maintain the adequate moisture level in the soil for the particular crop being farmed. The design is implemented using an Arduino microcontroller connected to a soil moisture sensor, a relay to control the water pump, as well as a GSM module to send data to a remote server. The remote server is used to represent data on the level of moisture in the soil to the farmers, based on the readings from the monitoring station.

Foreword

Sensors and actuators are bringing the Internet of Things to life in a multitude of fields, from agriculture to smart cities. Devices shifted from embedded to connected intelligence - at mist, fog, or cloud level – multiplying exciting opportunities in all aspects of our daily life.

The IoT-rush comes with its share of privacy breaches. While some corporations induce the number of household members based on smart door locks, others classify household activities based on smart appliances electricity logs.

How to harness the realm of possibilities of smart systems in an ethical and sustainable way?

Throughout these rich and relevant research articles, carefully curated by IoT experts Dr. Bhavya Alankar, Dr. Harleen Kaur and Dr. Ritu Chauhan, we find practical tools and directions for successfully transforming the Internet of Things in order to support next-generation smart systems.

The book is a must read for IoT engineers, designers, architects, as well as curious citizen and policy makers, in search for guidance on how to build a smart world in which people actually want to live.

Diana Derval
DervalResearch, China

Diana Derval, *PhD, EMBA, Chair and Research Director of DervalResearch, has pioneered research in human behavior and preferences. Harvard Business Review contributor, Jury for CES Asia Innovation Awards, patented inventor of the Hormonal Quotient® (HQ) nominated for the Edison Awards, member of the Society for Behavioral Neuroendocrinology, Author of the Derval Color Test® taken by 10+ million people, of the Iversity course on "The Science of Colors", and of the Springer Nature books "Designing Luxury Brands", "Hormones, Talent, and Career", and "The Right Sensory Mix"-finalist of the Berry-AMA Award for best marketing book and recommended by Philip Kotler- Diana turns fascinating neuroscientific breakthroughs into powerful business frameworks to decode and predict human traits, motivations, and behavior. Diana Derval has accelerated the development of Fortune 500 firms including Sephora, Michelin, Sofitel, Philips, and L'Oréal, and teaches Innovation, Luxury, Design Thinking and Neurosciences at Sorbonne Business School, IFA, Fudan/MIT and Donghua University. Over 25,000 professionals enjoyed her inspirational lectures, TEDx talks, and workshops from Paris to Shanghai.*

Preface

THE CONTEXT

Recent, era has witnessed a wide appraisal in IT based technology which has irrationally changed the overall lifestyle of human race. Moreover, the cities have intervened into smart cities where the global impact of Internet of Things (IoT) had undoubtedly transformed the varied application domains. IoT have evolved itself for real world requirements, which have resulted in total automation in several operations. However, IoT is massive technological intervention which has wide recognition around the globe. All leading organization are adopting new technology to automate itself from upcoming trends.

The internet of things (IoT) has massive potential to transform current business models and enhance human lifestyles. With the current pace of research, IoT will soon find many new horizons to touch. IoT is now providing a base of technological advancement in various realms such as pervasive healthcare, smart homes, smart cities, connected logistics, automated supply chain, manufacturing units, and many more. IoT is also paving the path for the emergence of the digital revolution in industrial technology, termed Industry 4.0.

Transforming the Internet of Things for Next-Generation Smart Systems focuses on the internet of things (IoT) and how it is involved in modern day technologies in a variety of domains. The multi-disciplinary view of IoT provided within this book makes it an ideal reference work for IT specialists, technologists, engineers, developers, practitioners, researchers, academicians, and students interested in how IoT will be implemented in the next generation of smart systems and play an integral role in advancing technology in the future.

WHAT THIS BOOK OFFERS

This book gives the insight of IoT and transformation of adaptive technology into smart system. It intends to globally understand the necessity or trend of IoT with new intervention in varied application domains which can forfeited benefit the researchers and scientists around the globe with research impact. We hope, this book will draw and illustrate the contribution of varied authors which can relatively be proceeded with more acknowledged research.

Bhavya Alankar
Jamia Hamdard, India

Harleen Kaur
Hamdard University, India

Ritu Chauhan
Amity University, India

Acknowledgment

We would like to acknowledge the role of the National Council for Science and Technology Communication (NCSTC), Department of Science & Technology (DST), Ministry of Science and Technology, Govt. of India, India. This work is catalyzed and supported by the research grant funded by the National Council for Science and Technology Communication (NCSTC), DST, New Delhi, India.

Chapter 1
Unleashing the Convergence of Cloud Computing With Internet of Things:
Drivers for Integration

Nalina Suresh
University of Namibia, Namibia

Valerianus Hashiyana
University of Namibia, Namibia

Martin Mabeifam Ujakpa
International University of Management, Namibia

Anton Limbo
University of Namibia, Namibia

Gloria E. Iyawa
Namibia University of Science and Technology, Namibia

Ntinda Maria Ndapewa
University of Namibia, Namibia

ABSTRACT

The term "cloud of things" is currently in the forefront of computer research topics due to its vital role towards the internet of things. To integrate cloud computing and internet of things into a single technology or worldview, this chapter discussed the evolution of internet of things and cloud computing and reviewed literature on cloud computing and internet of things and their possible integration. The chapter also discussed the importance of cloud internet of things, its architecture, and operation; the need to integrate internet of things and cloud computing; and challenges of cloud internet of things. The chapter then used the identified open issues and future direction to propose a generic cloud internet of things architecture and pilot simulated the results to verify the possibility and effectiveness of cloud computing and internet of things (CIoTs) integration. The researchers believe that the chapter will provide a better insight for anyone who wishes to carry out research in the field of cloud internet of things.

DOI: 10.4018/978-1-7998-7541-3.ch001

INTRODUCTION

By definition, cloud computing is an on demand service model for the provision of Information Technology. Virtualization and distributed computing technologies forms its foundation. Further it has the technological ability of transferring terabytes of torrents between data centers. Internet of Things (IoT) is defined as a network of physical objects, and devices that contain embedded technology (like intelligent sensors, controllers etc.) which can communicate, sense, or interact with internal or external systems. With IoTs, any physical object or devices; virtually 'anything' can become part of IoTs based services; and by this, a lot of data, generated. In order to create more valuable services, the data generated need to be managed based on its requirements. The Cloud Computing of IoT (CIoT) can be viewed as an advanced technology that hinges on the named pillars below (Rao et al., 2012).

1.   Anything is identifiable at Anytime and Anywhere
2.   Anything can communicate at Anytime and Anywhere
3.   Anything can interact Anywhere and Anytime

Integration of IoTs with the Cloud is becoming a novel technological approach and is termed as Cloud of Things (CoTs). IoTs and Wireless Sensor Networks (WSNs) seem to have had an ever increasing data and also related underlying resources, to which CoTs, serve as the means of handling them. IoTs and CoTs inadvertently rely on WSNs reception to the extent that, WSNs have increasingly become unavoidable, and making it a vital segment of the future Internet.

Characterized by real world things with limitation in storage and processing capacity, IoTs generally have the consequential issues of security, privacy, reliability and performance,. At the same time, cloud computing has virtually unlimited capabilities such as storage, processing power, privacy and security. Cloud computing provides infinite computation and storage through a shared pool of resources, which can dynamically be allocated and easily used by any IoT application.

The Internet of Things (IoT) worldview depends on smart and self-arranging hubs (things), interconnected in a dynamic and worldwide system framework. The web of things generally alludes to this present reality and seemingly insignificant details restricted capacity and the vital issues about unwavering quality, execution, security and protection. Like Cloud computing, distributed computing practically has boundless limit of capacity and handling power which can be retailored or further developed innovatively to any rate to a specific degree to handle the vast majority of the Internet of things (Chen et al., 2014; Tao, 2014; Wang et al., 2014) restricted storage and processing capacity. In this manner, a novel IT worldview in which Cloud computing and IoT, are integrally consolidated, is relied upon in the present and the future and in this study, the researchers call it, Cloud IoT (CIoT).

Cloud computing and IoT serve to increase efficiency in our everyday tasks, and the two have a complimentary relationship. IoT generates massive amounts of big data, and Cloud computing provides a pathway for the generated big data to travel through to its destination. Amazon Web Services (AWS), one of the several IoT Cloud platforms at work today, points out six advantages and benefits of cloud computing as follows:

- Enables a user to only pay for computing resources that they use.
- Enable service providers (such as AWS) to achieve greater economies of scale: thus reduced service costs for customers.

- Takes of Infrastructure capacity needs and maintenance, from users shoulder and thus enabling users to deploy any application anywhere in the world within minutes
- Ensures the availability of resources with increased speed and agility to users (such as developers).

To exploit the said advantages and benefits above and improve on the storage and processing capacity of IoTs, this paper attempted integrating Cloud computing and IoT, referred to hereafter as Cloud IoT (CIoT). This is necessary as the researchers foresee CIoTs playing crucial and vital roles in future. The researchers envisage facing some challenges to undertake this research. These challenges include the challenges that Parstream (2015) faced when he carried out similar research in which he collected data from IoT projects. The challenges that Parstream (2015) faced are illustrated in Figure 1.

Figure 1. Challenges in collecting and analyzing data from IoT projects (Source: Parstream (2015)

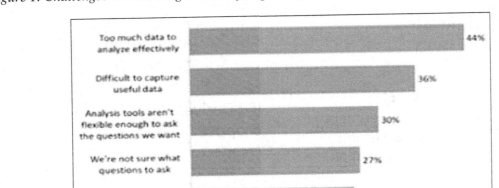

The Internet of Things (IoT) is starting to transform how we live our lives, but all of the added convenience and increased efficiency comes at a cost. The IoT is generating an unprecedented amount of data, which in turn puts a tremendous strain on the Internet infrastructure. As a result, companies are working to find ways to alleviate the said strain. Cloud computing will be a major part of that, especially by making all of the connected devices work together at an improved speed and storage capacity. But there are some significant differences between Cloud computing and the IoT and this will play out in the coming years as more and more data is generated.

RESEARCH METHODOLOGY

As a way of overcoming the challenges of collecting and analyzing IoT data as enumerated by Parstream (2015) in Figure 1 and to achieve the aim of integrating Cloud computing and IoTs, this paper followed applied the steps outlined in Figure 2 to carry out a quantitative and simulation experiment using NetLab

Figure 2. Quantitative and simulation experiment to integrating Cloud Computing and IoTs

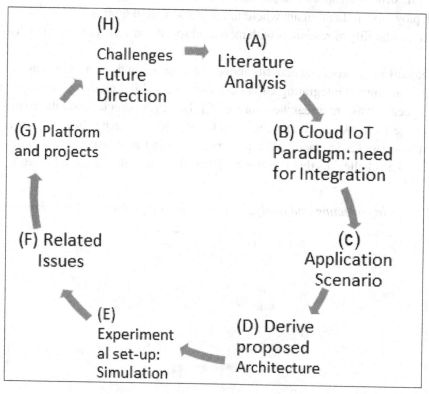

1. Firstly literature review is undertaken to unearth themes of common interest of Cloud computing, IoT, CIoT and their paradigms.
2. CIoT paradigm is discussed to highlight the complementarity and necessity for their integration.
3. The new application scenarios emanating from the acceptance of CIoT paradigm, is then detailed.
4. Fourth, researchers proposed an architecture for CIoT.
5. Fifth, the researchers simulated the CIoT using NetLab toolkit for interaction among physical and digital objects with several CIoT services.
6. Six, the researchers analyzed the CIoT paradigm and the application scenarios.
7. Seven, the researchers describe the main platforms and research projects in the field of CIoT.
8. Finally, the researchers derive the open issues and future directions in the field of CIoT.

Drivers required for Integrating Cloud and IoT were deduced from the document reviews which resulted in the architectural design in Figure 2 in the subsection titled, CIoT. Through simulation, the researchers then demonstrated the CIoT integration. This includes determining the necessity of CIoT architecture. This research provides a proposed recommendation from the technological aspects.

Literature Review

The two universes of Cloud computing and IoT have seen a quick and autonomous development. These universes differ from each other and, stunningly better, their qualities in a regularly correlative and augmentary manner. In totality, IoT can profit purposely from the boundless abilities and assets of Cloud computing to make up for its innovative limitations (e.g., capacity, preparing, and correspondence). For instance, the Cloud can offer powerful answers to the shortfall of IoT to enhance its administration, structure and applications including the things or the information created by the application (Lee et al., 2010). The Cloud can enhance IoT by extending its degree to manage certifiable 'things' in a more co-ordinated disseminated way. Most of the time, the Cloud is able give a middle layer between the 'things' and the applications: and hence concealing the unpredictability and functionalities that is important to enable the actualization of the applications. This affects future application advancement, where data social event, preparing, and transmission will produce new difficulties, particularly in a multi-cloud environment (European Commission, 2013).

IoT assets include capacity assets (European Commission, 2013) (Rao et al., 2012), computational assets (Rao et al., 2012) (Zaslavsky et al., 2013) and correspondence assets (Zaslavsky et al., 2013) (Rao et al., 2012) (Parwekar, 2011). Capacity assets: by definition, include lots of data sources/things, which eventually results in massive measure of non-organized or / and semi-organized information (European Commission, 2013). The massive non-organized or / and semi-organized information usually have three regular attributes of Big Data (Zikopoulos & Eaton, 2011): volume (i.e., information estimate), assortment (i.e., information sorts), and speed (i.e., information era recurrence). Subsequently it suggests gathering, getting to, preparing, imagining, documenting, sharing, and looking for a lot of information (Rao et al., 2012). Offering essentially boundless, ease, and on-request stockpiling limit, Cloud computing is the most advantageous and practical answer to manage information delivered by IoT (Rao et al., 2012). This incorporation understands another merging situation (European Commission, 2013), where new open doors emerge for information accumulation (Fox et al., 2012), joining (Dash et al., 2010), and passing onto outsiders. Once into the Cloud, information can be dealt with in a homogeneous way through the standard APIs (Fox et al., 2012), applying top-level security to ensure (Zaslavsky et al., 2013), and specifically got to and pictured from wherever (Rao et al., 2012).

Computational assets: IoT gadgets have constrained preparing assets that don't permit nearby in-formation handling. Information gathered is normally transmitted to all the more intense hubs where preparing is conceivable, yet adaptability is trying to accomplish without an appropriate foundation. The boundless handling capacities of the Cloud and its on-request display to permit IoT preparing, should be appropriately fulfilled and empower investigations of uncommon multifaceted nature (). Information driven basic leadership and forecast calculations would be conceivable easily and would give expanding incomes and lessened dangers (Parwekar, 2011). Different points of view is perform constant prepar-ing (on-the-fly), to actualize adaptable, continuous, cooperative, sensor-driven applications, to oversee complex occasions, and to execute assignment offloading for vitality sparing.

Correspondence assets: One of the prerequisites of IoT is to make IP-empowered gadgets communicate through devoted equipment, and the support for such correspondence can be extremely costly. Cloud offers a compelling but shabby answer for interfacing, tracking, and overseeing anything from anyplace whenever utilizing altered entrances and worked as a part of the applications. On the account of the ac-cessibility of fast systems, it empowers the checking and control of remote 'things', their coordination, their interchanges, and the constant access to the delivered information (Parwekar, 2011).

Necessaries and Complementarities for Integration

The convergence of the Cloud and IoT have had fair developments thus far. As mentioned earlier and for emphasis, the current status quo of the IoTs in terms of storage capacity and processing power, can be enhanced greatly by the cloud's practical boundless limit and assets to compensate for the IoT specialized requirements. (e.g., capacity, preparing, and vitality). Particularly, the Cloud can offer a viable potential for the execution of IoT administration, organization and applications.

The correlative attributes of the Cloud and IoT emerging from diverse recommendations in literature, has motivated the CIoT, as proposed in this study. The integral qualities of distributed computing and Internet of things is alluring a direct result of the diverse recommendations reported in writing and empowering CIoT worldview as depicted in Table 1. Principally, the Cloud serves as a layer between the 'things' and the applications and the layer, holds all the intricacies and functionalities necessary to implement the latter. The complementary characteristics of Cloud computing and IoTs is attractive because of the different proposals reported in literature and encouraging CIoT paradigm. Table 1 outlines Cloud and IoT complementarity and Integration.

One of the necessities of the IoTs is to make IPs get to gadgets by communicating through the devoted equipment, and bolstering the correspondence, which can be exceptionally costly. Cloud association gives a successful and modest arrangement, for example, following and overseeing anything whenever from wherever to utilize a custom entry and implicit application(s).

Table 1. Complementarities and Integration of Issues

Internet of Things	Cloud Computing
pervasive	ubiquitous
real world	virtual resources
limited computational	unlimited computational
limited storage	unlimited storage
point of convergence	service delivery
big data source	means to manage big data

(Source: Alessio, De-Donato, Persico, and Pescapé, 2014)

CIoT Application Driven Scenarios

IoT delivers a lot of unstructured or semi-organized information that have the three noteworthy information attributes: volume, speed and assortment. Consequently this points toward the accumulation, procurement, preparing and perception, share and seeking of lots of information. With computation, connectivity, and data storage becoming more advanced and universal, there has been an explosion of IoT based application solutions in diversified domains from health care to public safety, assembly line scheduling to manufacturing and various other technological domains (Rao et al., 2012). Given its practical boundless and on-request stockpiling limit, minimal effort, the cloud is the most helpful and financially savvy answers to manage the information produced by IoT (Rao et al., 2012). With IoT, Information driven leadership and forecast calculations would be conceivable: hence requiring little to no effort and thus leading to expanded income net and reduced danger. The combination of IoT with the Cloud drives the above to possibilities. Further, it makes it possible for extra elements creation and utilization, improved usability, and reduced costs. Extension of the cloud through IoTs outlined in Table 2.

Table 2. Extension of Cloud through the IoTs

SaaS	Sensing as a Service	Providing ubiquitous access to sensor data.
SAaaS	Sensig and Actuation as a Service	Enabling automatic control logics implemented in the cloud.
SEaaS	Sensor Event as a Service	Dispatching messaging services triggered by sensor events.
SenaaS	Sensor as a Service	Enabling ubiquitous management of remote sensors.
DBaaS	DataBase as a Service	Enabling ubiquitous database management.
DaaS	Data as a Service	Providing ubiquitous access to any kind of data.
EaaS	Ethernet as a Service	Providing ubiquitous layer-2 connectivity to remote devices.
IPMaas	Identity and Policy Management as a Service	Enabling ubiquitous access to policy and identity management functionalities.
VSaaS	Video Surveillance as a Service	Providing ubiquitous access to recorded video and implementing complex analyses in the cloud.

(Source: Rao, Saluia, Sharma, Mittal, and Sharma, 2012; Suciu, Vulpe, Halunga, Fratu, Todoran, and Suciu, 2013)

Figure 3. Application Scenarios of CIoT

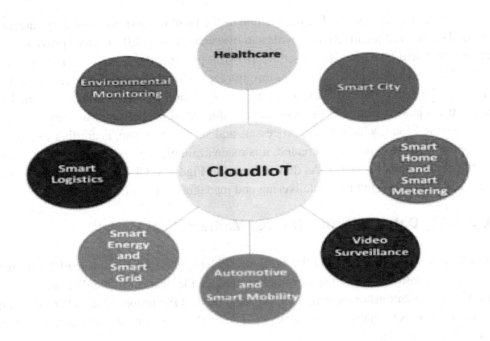

The above Figure 3 depicts some envisioned CIoT application scenarios which are discussed further as below:

- Healthcare. In this scenario, Cloud acceptance usually lead to the abstraction of technical details, removing the necessity for know-how in, or control over, the technology infrastructure (Lohr et al., 2010), (Doukas & Maglogiannis, 2012), and it represents a promising solution for managing healthcare sensor data efficiently (Mitton et al., 2012).
- Smart City. Some recently suggested solutions indicate the use of Cloud architectures to enable the detection, connection, and incorporation of sensors and actuators that ends in platforms with the ability to provide and support ubiquitous connectivity and real-time smart cities applications (Kamilaris et al., 2011).
- Smart Home and Smart Metering. IoT has large application in home environments, where heterogeneous embedded devices enable the automation of common in-house activities. In this scenario, the Cloud is the best candidate for building flexible applications with only a few lines of code, making home automation a trivial task (Koubaa et al., 2016).
- Video Surveillance. This video is a tool of the greatest importance as it is applicable to several security-related applications.
- Automotive and Smart Mobility. IoT is expected to offer promising solutions to transform transportation and automobile systems and services.
- Smart Energy and Smart Grid. IoT and Cloud can be effectively merged to provide intelligent management of energy distribution and consumption in both local and wide area heterogeneous environments.

Generic CIoT Architecture and Description

The three-Tier architecture depicted in Figure 4 is one of the form to solve IoT issues by integrating it with the Cloud. This layered architecture provides an overview of how IoT services provisioning IoT-cloud communication will take place (Chen et al., 2014): it presents the overall CIoT communication pattern. Various IoTs generate data, which passes through each of the layers presented in Figure 4 (Wu et al., 2015). Further, the data is transmitted through a communication channel where a lot of multimedia communication takes place. A number of examples are illustrated in the layers in Figure 4. The data ultimately reaches the Cloud, which stores, processes, and secures the data, according to the requirements of the service. Once the service is created, it is made available to the end user, which resides the other end of the cloud: the access layer. As demonstrated in Figure 4, CIoT plays an important role in the processes above, including but not to delivering and managing services.

Structure Fidelity Data Collection (SFDC) Mathematical Model

Moving forward, the researchers adopted the Structure Fidelity Data Collection (SFDC) framework, leveraging on the correlations between sensor nodes to reduce the distortion of collected data (Netlab-toolkit, n.d.) during simulations or experiments. Figure 5 depicts the proposed data flow diagram and interaction among the IoTs with thee Cloud services. The mathematical model and analysis associated with the Figure 5 is as follows:

Figure 4. Generic CIoT Architecture

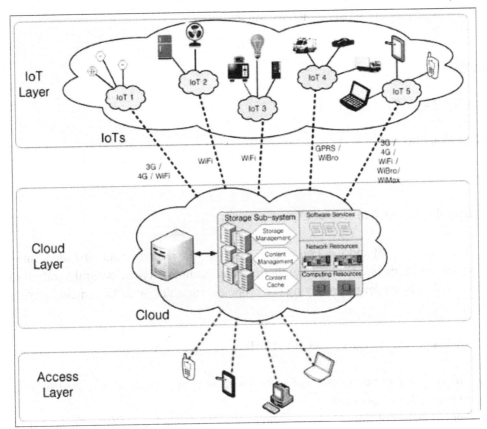

Figure 5. Proposed Data Flow Diagram

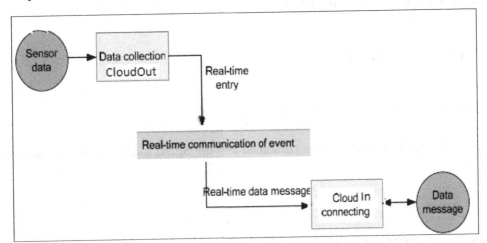

If the Sensor data is $S_d = \{v_1, v_2, v_3, v_4, \ldots\ldots v_n\}$
Where, $v_1, v_2, v_3, v_4, \ldots\ldots v$ are sensor generated values
Then,

CloudOut(COt) $\in S_d$

COt $= \forall\, y \in S_d: \exists\, x \in : $ CIoT (x, y)

Similarly, CloudIn(CIn) $\subset S_d$

CIn $\rightarrow$ COt

"CIn if and only if COt", i,e CIn <==> COt

Let $x = \{x_i | i = 1, 2, \cdots, N\}$ and $y = \{y_i | i = 1, 2, \cdots, N\}$ be the reference and distorted sample values, where i is the sample index and N is the number of value samples. The structural similarity index between values x and y can be defined applying our proposed variables on SFDC equation (Netlabtoolkit, n.d.) as below:

$$\text{SFDC}(x, y) = \text{CIoT}_{mean}\,(x, y) \cdot \text{CIoT}_{Stdev}\,(x, y) \cdot \text{CIoTco-variance}\,(x, y)\ldots\ldots\ldots \tag{1}$$

Let $\mu_x, \mu_y, \sigma^2_x, \sigma^2_y$ and σ_{xy} be the mean of x, the mean of y, the variance of x, the variance of y, and the covariance of x and y, respectively

The covariance can be considered as a measurement of how much one COt value is changed nonlinearly to the other CIn value being compared.

$$\textbf{CIoT}_{mean}\,(\textbf{x, y}) = 2\,\mu_x\,\mu_y + C'_1 / \mu^2_x\,\mu^2_y + C'_1 \ldots \tag{2}$$

$$\textbf{CIoT}_{Stdev}\,(\textbf{x, y}) = 2\,\sigma_x\,\sigma_y + C_2 / \sigma^2_x\,\sigma^2_y + C'_2 \ldots \tag{3}$$

$$\textbf{CIoTco-variance}\,(\textbf{x, y}) = \sigma_{xy} + C_3 / \sigma_x\,\sigma_y + C'_3 \ldots \tag{4}$$

The equations (1), (2), (3) and (4) are combined and the SFDC index (Netlabtoolkit, n.d.) can be rewritten by:

$$\text{SFDC}(x,y) = \ldots\ldots \tag{5}$$

Based on the equation (5)[REMOVED HYPERLINK FIELD], the SFDC index has the following three properties:

4. Symmetry: *SFDC* $(x, y) = $ *SFDC* (y, x).

5. Boundedness: *SFDC* $(x, y) \in [-1, 1]$ since the three components of Equation (1) range from [0,1], [0,1] and [−1,1], respectively.
6. Unique maximum: *SFDC* $(x, y) = 1$ if and only if $x = y$ (in discrete representations, $x_i = y_i$ for all
7. $i = 1, \cdots, N$).

Simulation Setup

This study considered a CIoT scenario by undertaking a simulation. The objective of the simulation was to test the performance of the proposed CIoT (Throughput). The snap shots of the simulation results are presented below. Using the existing Cloud services, organisations have developed toolkits for their integration in CIoT frameworks. For this paper, the researchers demonstrated CIoT integration using NetLab. Netlab toolkit allows for interaction between physical and digital objects (e.g. controlling video movies through Arduino). NetLab's two widgets, CouldIn and CloudOut, allows for interaction with several CloudIoT services. This includes periodical sending of data from 'things' to Cloud services or retrieving data from the said services.

Figure 6 to 20 are screen shots of the simulation results. The demonstration of the Internet of Things CloudIn/Out Widgets are described in the next paragraphs. The Widgets send data: CloudOut sends data up to Cloud and CloudIn retrieve's the values from Cloud back down. As seen towards the left in the Figure 6, Analogin Widget connected to Arduino with an audio attached so that is sending the data from the Arduino up to the Cloud every 2sec. Also, it is retrieving that data stream from Cloud also every 2sec. continuously, values are retrieved through the Cloudin that triggers the audio. As soon as the values synchronize, the audio stop's as shown in Figures 6 and 7. The Steps used for the simulation setup are listed below:

Figure 6. CloudIn/Out Widgets setup

Figure 7. CloudIn/Out Widgets setup

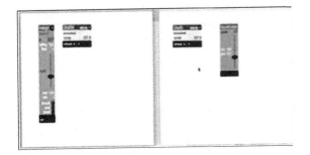

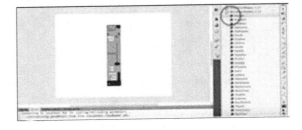

Step 1

-   Create a new doc: the researchers Put Analogin on the stage and named that as input 0.
-   Put a CloudOut: the researchers sent that feed from Arduino up to the cloud
-   Figure 8 shows the sensing device, in this case Arduino configuration

Figure 8. Arduino setup for CloudOut Widgets

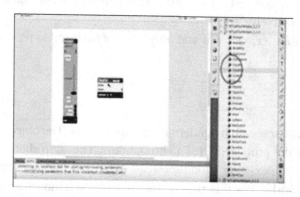

Step 2

Properties of CloudOut that should be noted here areAPI key, Channel and Datafeed. The three Cloud Services that NetLab support are open.Sense, thinkspeak service provider (makes the Iobridge devcies) and COSM. For this study, the researchers used COSM in the simulation as shown in the Figures 9 to 12.

Figure 9. COSM cloud services setup for API key

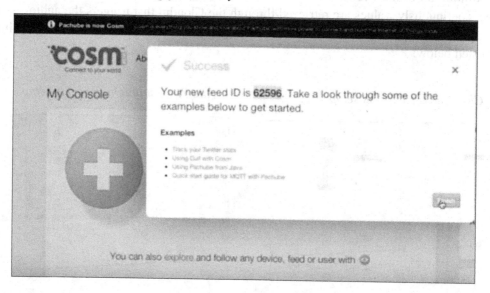

Step 3

- Create a device or channel: the researchers called it as Audrino.
- Give it some Tags like Knob and Auridon.
- Create the feed. And notice the id. the researchers needed to put the id to the channel
- Add data feed, what COSM causes data stream. Name Analogin0.
- Notice no data created yet

Figure 10. COSM Cloud services setup for API key

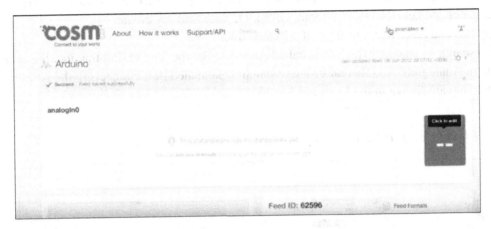

Figure 11. COSM Cloud setup for services

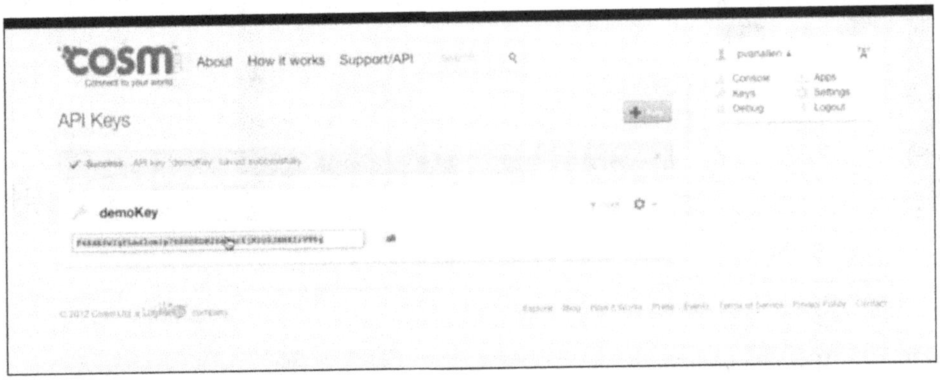

Figure 12. COSM Cloud setup for services

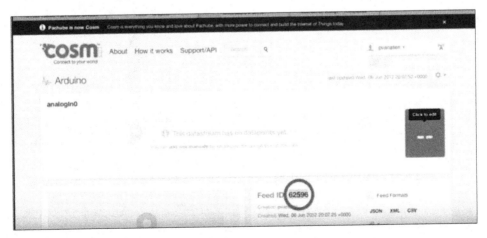

- The last thing is create a key like a pw for accessing your feed whether reading or writing
- And say create key. Call this as demo. Give all rights. And say create.
- Copy this and paste that into flash of API key area.
- Get researchers channel which they called Arduino. Get the feed id. Put that into the channel.
- Type in Analogin0. And thereafter the researchers launched it, it connects to the Arduino.
- These step are shown in the Figures 13 to 15.

Figure 13. CloudIn Widgets launch

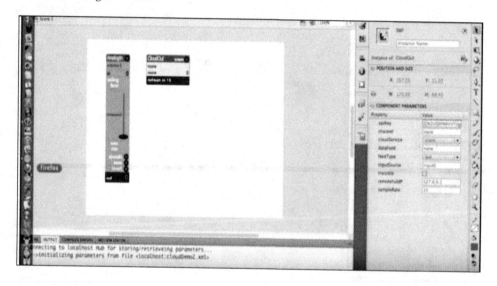

Figure 14. CloudIn Widgets launch

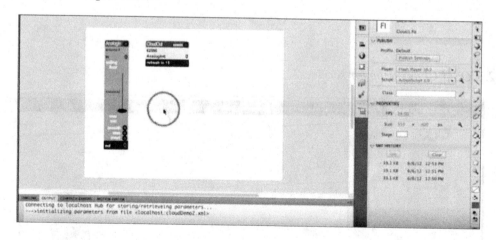

Figure 15. CloudIn Widgets launch

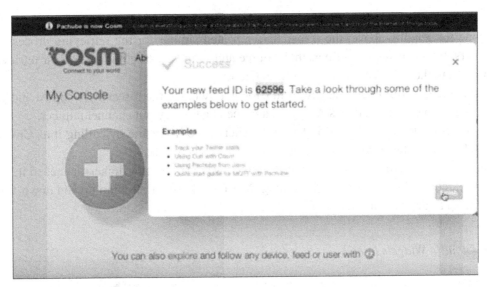

Step 4

- The researchers connected to the COSM services, which only sent values periodically. By default it sent values every 13secs.
- The researchers set it to send and retrieve data from the services. With an Adjusted sample rate 2 sec, it sent every 2sec.
- Looking at feedtype, the researchers took an average of all values that were retrieved from the Widget from Arduino during the period of the 2sec. On this basis, it counted down and fired every 2 sec and sent that value up to the Cloud as shown in Figure 16.

Figure 16. Arduino and CloudOut streaming

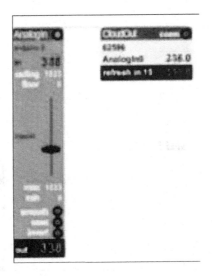

Step 5

- As done for sending above, the researchers repeated similar for the retrieving process. The researchers did this using CloudIn as their source and sound control Widget as their retrieval and it responded to the values coming down from Cloud.
- Going to the CloudIn to get the key from CloudOut widget, the researchers copied and pasted in and thusthat their data feed was Anlogin0 and then again they put channel number.
- Need to set cloud service COSM. The researchers gave it a name by calling it as Cloud b. The researchers the listened to the sound control as Cloud feedback.
- Connecting to CASM service, to last value which was 146, the researchers launched it.
- The researchers observed that both running, Move the knob of Arduino up and down, the values showed up as depicted in Figures 17 to 20.

Figure 17. CloudIn Widgets launch

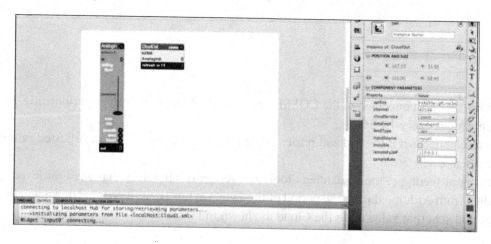

Figure 18. CloudIn Widgets launch

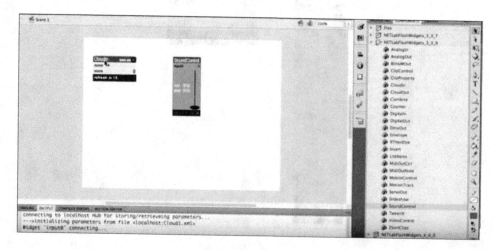

Figure 19. CloudIn Widgets launch

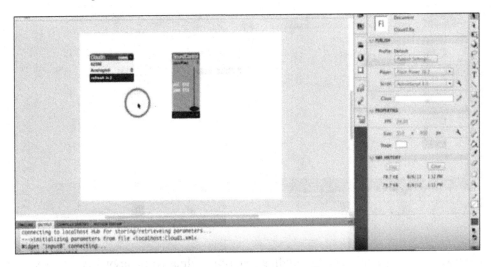

Figure 20. CloudIn Widgets launch

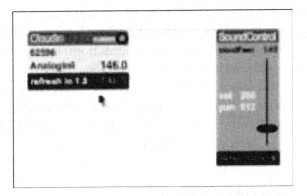

Step 6

8. The process goes to retrieve and sound starts when the sound synchronize audio stops. This does not happen immediately as it takes a couple of sec to send and receive and how they match up. This is illustrated in Figure 21.

RESULTS AND DISCUSSIONS

The result of the study show that sending and receiving data steam's from sensors to CloudIn/Out (CIn and COt) correlating to the SFDC equation. According to equation5 (i & iii):

9. When SFDC(x, y) = 1; That is, when sending (COt) and receiving (CIn) data messages synchronize as shown in Figure 20, it proves the efficient throughput of IoT.

Figure 21. CloudIn/Out Widgets launch

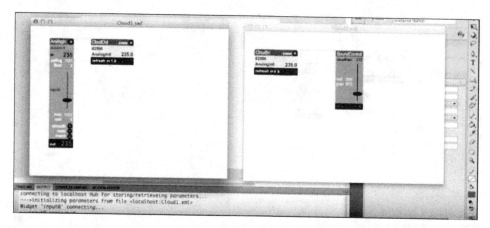

In this context, the study simulated by sending real time values from anywhere in the world and that was similarly retrieved from anywhere. It is completely asynchronous and feeds can be distributed and used anywhere in the world.

The CloudOut and CloudIn widgets sent and received data from open IoT (Internet of Things) Cloud services including Xively (formerly COSM and Pachube), Open.Sense, and ThingSpeak. These services allow for devices, software, or people to post and receive feeds from anywhere in the world. For example, a gesture sensor in Australia could send its values to the cloud, and video displays in London and Beijing can access the latest value of that sensor, and play different scenes based on the Australia sensor value. The key thing about these services is that feeds can be posted and retrieved from anywhere, without the need to have a direct connection between the sender and receiver. The IoT cloud service handles the asynchronous reception, storage, and delivery of the feeds.

RELATED ISSUES/LIMITATIONS OF IoT

In order to retrieve optimal results and proper functioning of the CIoT network, naïve software have to be developed to work with hardware components used in intelligent devices in IoT. Further, specific technology platforms are to be developed which will provide such kind of functionality to CIoT and add more intelligence to the existing network. The traditional algorithms used in cloud computing have to be modified and enhanced in order to make them compatible with the CIoT environment.

Consequently, the complex scenario of CIoT includes many aspects related to several varied topics, each of them imposing challenges when requiring specific capabilities to be satisfied. For instance, the following capabilities are required to guarantee trusted and efficient services: security, privacy, reliability, scalability, availability, portability, interoperability, pervasiveness, energy saving, cost effectiveness, seamlessly integrate them, and protocols that facilitate big data streaming from IoT to the cloud, QoS and QoE to mention a few (Rao et al., 2012)

CHALLENGES AND FUTURE DIRECTIONS

Some prime open challenges are discussed based on the IoT elements presented earlier. The challenges include IoT specific challenges such as privacy, participatory sensing, data analytics, energy efficiency, security, protocols, and Quality of Service.

A future roadmap of key developments in IoT research in the context of pervasive applications is shown in Figure 22, which includes the technology drivers and key application outcomes expected in the next decad. A CIoT can further extend an application as a service. Additionally, there is the need to address the challenges of the wide spectrum of CIoT applications.

Figure 22. Future Roadmap of key technological developments in the context of IoT application domains envisioned

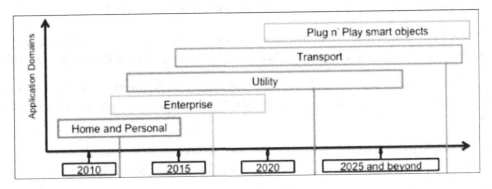

CONCLUSION

The current trend of using IoTs in isolation has a lot of limitations: among these limitations are; storage capacity challenge, computational power limitation, and security and lack of standard IoT framework. As suggested in this study, the integration of Cloud Computing and Internet of Things is the next future of the Internet platform. The emerging concept from the integration, CIoT, opens up a new stimulating paradigm for business and research. In this paper, the researchers worked at the specific goal of looking into the correlative of the Cloud and IoT and the primary drivers for converging them asone new domain, CIoT. Through the research, the researchers were able to unearth some details on the integration of Cloud Computing and Internet of Things in the perspective of applications, issues and challenges in IoT.

Considering that CIoT worldview brings a new dimension to IoT and empowers new applications, it is the hope of the researchers that, this study is useful to researchers and practitioners in the field of Cloud computing and IoT. Further, this study highlighted the huge potential of IoT, Cloud computing, their diversified engineering applications and their major issues which could be investigated in future research and hence narrowing down the gap of edging the CIoT concept closer to reality.

REFERENCES

Aazam, M., Huh, E. N., St-Hilaire, M., Lung, C. H., & Lambadaris, I. (2016). Cloud of Things: Integration of IoT with Cloud Computing. In A. Koubaa & E. Shakshuki (Eds.), *Robots and Sensor Clouds. Studies in Systems, Decision and Control* (Vol. 36). Springer. doi:10.1007/978-3-319-22168-7_4

Alagoz, F. (2010). From cloud computing to mobile Internet, from user focus to culture and hedonism: the crucible of mobile health care and wellness applications. In *ICPCA 2010*. IEEE. doi:10.1109/ICPCA.2010.5704072

Atzori, L., Iera, A., & Morabito, G. (2010). The Internet of Things: A survey. *Computer Networks*, *54*(15), 2787–2805. doi:10.1016/j.comnet.2010.05.010

Biswas, A. R., & Giaffreda, R. (2014). IoT and cloud convergence: Opportunities and challenges. *Internet of Things (WF-IoT), IEEE World Forum on*. 10.1109/WF-IoT.2014.6803194

Chen, Y., Zhao, S., & Zhai, Y. (2014). Construction of intelligent logistics system by RFID of Internet of things based on cloud computing. *Journal of Chemical and Pharmaceutical Research*, *6*(7), 1676–1679.

Dash, S. K., Mohapatra, S., & Pattnaik, P. K. (2010). A Survey on Application of Wireless Sensor Network Using Cloud Computing. *International Journal of Computer science & Engineering Technologies*, *1*(4), 50–55.

Doukas, C., & Maglogiannis, I. (2012). Bringing iot and cloud computing towards pervasive healthcare. In *Innovative Mobile and Internet Services in Ubiquitous Computing (IMIS), Sixth International Conference on*, (pp. 922–926). IEEE. 10.1109/IMIS.2012.26

European Commission. (2013). *Definition of a research and innovation policy leveraging Cloud Computing and IoT combination*. Tender specifications, SMART 2013/0037.

Fox, G. C., Kamburugamuve, S., & Hartman, R. D. (2012). Architecture and measured characteristics of a cloud based internet of things. In *Collaboration Technologies and Systems (CTS), International Conference*, (pp. 6–12). IEEE. 10.1109/CTS.2012.6261020

Gebremeskel, G. B., Chai, Y., & Yang, Z. (2014). The Paradigm of Big Data for Augmenting Internet of Vehicle into the Intelligent Cloud Computing Systems. In *Internet of Vehicles–Technologies and Services* (pp. 247–261). Springer International Publishing. doi:10.1007/978-3-319-11167-4_25

Gubbi, J., Buyya, R., Marusic, S., & Palaniswami, M. (2013). Internet of Things (IoT): A vision, architectural elements, and future directions. *Future Generation Computer Systems*, *29*(7), 1645–166. doi:10.1016/j.future.2013.01.010

Kamilaris, A., Pitsillides, A., & Trifa, V. (2011). The smart home meets the web of things. *International Journal of Ad Hoc and Ubiquitous Computing*, *7*(3), 145. doi:10.1504/IJAHUC.2011.040115

Kang, J., Yin, S., & Meng, W. (2014). An Intelligent Storage Management System Based on Cloud Computing and Internet of Things. In *Proceedings of International Conference on Computer Science and Information Technology*. Springer India, 10.1007/978-81-322-1759-6_57

Lee, K., Murray, D., Hughes, D., & Joosen, W. (2010). Extending sensor networks into the cloud using Amazon web services. *Networked Embedded Systems for Enterprise Applications (NESEA), IEEE International Conference.*

Lohr, Sadeghi, & Winandy. (2010). Securing the e-health cloud. In *Proceedings of the 1st ACM International Health Informatics Symposium*, (pp. 220–229). ACM.

Mitton, N., Papavassiliou, S., Puliafito, A., & Trivedi, K. S. (2012). Combining Cloud and sensors in a smart city environment. *EURASIP Journal on Wireless Communications and Networking, 2012*(1), 1–10. doi:10.1186/1687-1499-2012-247

Netlabtoolkit. (n.d.). https://www.netlabtoolkit.org/learning/tutorials/iot-cloud-services

Parwekar, P. (2011). From Internet of Things towards cloud of things. In *Computer and Communication Technology (ICCCT), 2nd International Conference on*, (pp. 329–333). IEEE. 10.1109/ICCCT.2011.6075156

Rao, B. P., Saluia, P., Sharma, N., Mittal, A., & Sharma, S. V. (2012). Cloud computing for Internet of Things & sensing based applications. In *Sensing Technology (ICST), Sixth International Conference*, (pp. 374–380). IEEE. 10.1109/ICSensT.2012.6461705

Suciu, G., Vulpe, A., Halunga, S., Fratu, O., Todoran, G., & Suciu, V. (2013). Smart Cities Built on Resilient Cloud Computing and Secure Internet of Things. In *Control Systems and Computer Science (CSCS), 19th International Conference on*, (pp. 513–518). IEEE. 10.1109/CSCS.2013.58

Tao, F. (2014). CCIoT-CMfg: cloud computing and Internet of Things based cloud manufacturing service system. Academic Press.

Vyas, Bhat, & Jha. (n.d.). IoT: Trends, Challenges and Future Scope. *International Journal of Computer Science & Communication, 7*(1), 186-197.

Wang, H. Z., Lin, G. W., Wang, J. Q., Gao, W. L., Chen, Y. F., & Duan, Q. L. (2014). Management of Big Data in the Internet of Things in Agriculture Based on Cloud Computing. *Applied Mechanics and Materials, 548*, 1438–1444. doi:10.4028/www.scientific.net/AMM.548-549.1438

Wu, M., Tan, L., & Xiong, N. (2015). A Structure Fidelity Approach for Big Data Collection in Wireless Sensor Networks. *Sensors (Basel), 15*(1), 248–273. doi:10.3390150100248 PMID:25609045

Zaslavsky, A., Perera, C., & Georgakopoulos, D. (2013). *Sensing as a service and big data.* arXiv preprint arXiv:1301.0159.

Zikopoulos, P., & Eaton, C. (2011). *Understanding big data: Analytics for enterprise class hadoop and streaming data.* McGraw-Hill Osborne Media.

KEY TERMS AND DEFINITIONS

Big Data: Is an extremely large amount and varied data sets that may be analyzed computationally to unearth valuable information patterns, trends, relations, and associations, especially relating to human treatments, behaviour and interactions that can help institutions make informed decisions.

Cloud Architecture: Is how individual technologies are integrated to create clouds—IT environments that abstract, pool, and share scalable resources across a network.

Cloud Computing: Is the delivery of on-demand computing services: from applications to storage and processing power – typically over the internet and on a pay-as-you-go basis.

Cloud of Things: It is a high-performance, cloud-based IoT application platform for the internet of things.

Embedded Device: Is a combination of a computer processor, computer memory, and input/output peripheral devices.

Healthcare: Is the organized provision of medical or patient care to individuals or a community.

Internet of Things: Is a system of interrelated computing devices, mechanical and digital machines, objects, animals or people that are provided with unique identifiers and the ability to transfer data over a network without requiring human-to-human or human-to-computer interaction.

Mobility: Is a user-centric concept – recognizing that transportation products and services must be responsive to the needs, habits, and preferences of consumers.

NetLab: Is an active member of a number of a number of international networks concerned with research or web archiving.

Smart City: Is an urban area that uses different types of electronic methods and sensors to collect data.

Smart Energy: Is the process of devices using energy-efficiency. It focuses on powerful, sustainable renewable **energy** sources that promote greater eco-friendliness while driving down costs.

Smart Home: Is a home with a system that connects with certain appliances by automating specific tasks.

Smart Meter: Is an electronic device that records information on the consumption of electric energy, voltage levels, current, and power factor.

Smart Mobility: It is a phenomenon that undoubtedly transform the future of transport, as people seek the simplest way to get to where they want to be.

Wireless Sensor Networks: Is a group of spatially dispersed and dedicated sensors for monitoring and recording the physical conditions of the environment and organizing the collected data at a central location.

Chapter 2
Apprising Trust Key Management in IoT Cross-Layer Framework

Rachna Jain
https://orcid.org/0000-0001-9794-614X
JSS Academy of Technical Education, Noida, India

ABSTRACT

Internet of things (IoT) networks is the buzzword these days in Industry 4.0. IoT nodes are resource constrained and should be light enough to minimize the power consumption. IoT paradigm does not depend on human intervention at each and every step. There is a need of "trust" between communicating entities. Devices at physical layer are vulnerable to various attacks such as denial of service (DoS) attack, wormhole attack, etc. Trust becomes more important when vulnerability of attacks increases to the devices. This establishment of trust helps in handling risks in a controlled way in unpredicted situations as well as providing better services at infrastructure level. Social environments can evaluate trust while seeing the relationship between interacting parties; however, in service-oriented industries quality of service (QoS) parameters must be maintained while evaluating trust. So, in this chapter a unique metric expected transmission count (ETX) is employed for implementing QoS while evaluating trust between interacting entities using Cooja simulator.

1. INTRODUCTION

IoT devices connect physical world with cyberspace through billion objects. Service Oriented Architecture (SOA) in large heterogeneous networks impart interoperability between devices (Chen, Guo and Bao, 2014). Figure 1 displays the enormous power held by IoT devices which is located in central cloud resulted in Mobility as a Service (MaaS) applications. User mobility data provides valuable information through mobile sensors using MaaS architecture. Mobility management attaches the mobile nodes for packets transmission to the serving networks (Chai, Choi and Jeong, 2015). IoT devices have opened a whole new range of applications in Industrial IoT networks (Da Xu et al., 2014). IoT devices have been

DOI: 10.4018/978-1-7998-7541-3.ch002

equipped with sensing and processing capabilities. Near Field Communication and Sensor networks have evolved along with IoT networks in industry 4.0 (Whitmore et al., 2015). Wireless 5G networks will further transform 5G-IoT networks to seamlessly integrate with other artificial intelligence operated devices (Li et al., 2018).

Figure 1. IoT devices along with Mobility as a Service (MaaS) applications

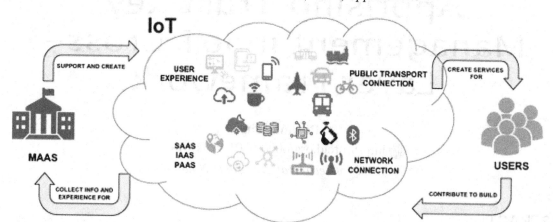

Different technologies have proffered the applications of IoT networks. IoT devices have helped in reducing or bridging gap between virtual and physical world. Scalabililty, interoperability and security are the key aspects to be pondered upon while framing an IoT network (Ulmer et al., 2013). Standardization, processing and storage of data and trust management are open challenges in IoT implementation (Mohanta et al., 2020). Another major task in heterogeneous networks like IoT is to enhance QoS of the network by allocating minimum resources (Abedin et al., 2018). CISCO has illustrated that 50 billion IoT devices will be connected by the end of 2020 (Mohanta et al., 2020). Figure 2 displays the evolution of IoT framework. IoT technology comprises of RFID sensing, smart cities, intelligence in objects (Yang et al., 2010). RFID reader and tags allow easy recognition of the objects. Due to its amazing benefits RFID systems are widely used in inventory management, production and access control.

2. SERVICE ORIENTED ARCHITECTURE (SOA)

Service Oriented Architecture (SOA) prevalent in IoT frameworks comprises of sensing layer, network layer, service layer and interface layer (Li, Xu and Zhao, 2015). Sensing layer comprises of sensing smart systems with RFID tags or Universal Unique Identifier (UUID) identities. The goal of devices should be to minimize the usage of resources like energy consumption. Another aim should be coexistence of different networking technologies such as Zigbee, Z-wave, WLAN and Bluetooth. Main function of network layer is interconnectedness of these devices and sharing of data among these devices. Their communication should enhance QoS of the network (Jain and Kashyap, 2019). Service layer provides middleware among participating technologies. The main activities of this layer are to initiate service discovery, composition, API and management of trust. Figure 3 shows the importance of trust in IoT framework. Interface layer provide the common standards such that different vendors can interact amongst themselves. Yang et al., 2017 have thoroughly analyzed the security issues for an IoT framework.

Figure 2. Growth of IoT framework
(Source: Li, S., Da Xu, L. and Zhao, S., 2015)

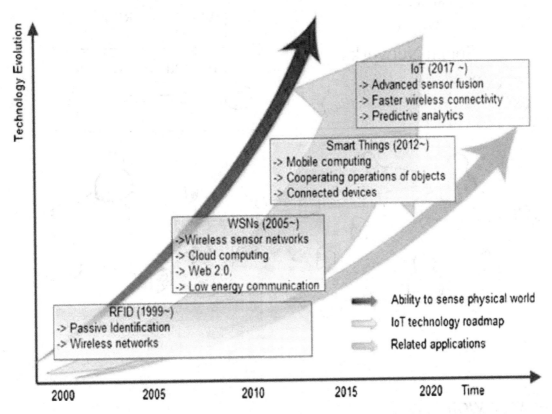

Figure 4 displays the extended framework of models which is further subdivided as composition based, propagation based, aggregation based, formation and layer based. For example propagation based trust could be achieved either in collective manner i.e. hub or spoke model or in distributed fashion such as used in Distributed Ledger Technology (DLT) (Ranathunga, Marfievici, McGibney and Rea, 2020). Trust aggregation techniques could be summarized either as fuzzy logic, bayesian networks, belief theory or machine learning based techniques.

SOA architecture adopted by IoT devices is prone to attacks such as wormhole attack, Denial of Service (DOS) attack, spam networks, sybill, falsification etc. Figure 5 shows the major attacks at different layers of IoT framework.

Figure 3. Trust Management in IoT framework
(Source: I.Chen, J. Guo, F. Bao, 2016)

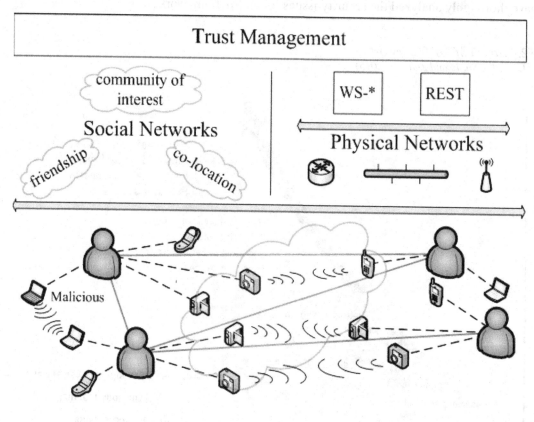

3. RELATED WORK

Systematic Literature Review (SLR) shows that IoT devices are having applications in all major domains such as healthcare, monitoring of environment as well as commercial aspects (Asghari, Rahmani and Javadi, 2019). IoT networks are helpful in establishing ubiquitous computing and Machine to Machine (M2M) interface (Li,Xu and Zao 2015). Quality of Service (QoS) issues such as energy consumption by nodes, attack vulnerabilities must be considered at length before implementing IoT networks. Devices connected must be energy efficient and secure. It has been elucidated that adaptive, trust managed IoT devices handling using adaptive filtering. In the proposed approach to save the scared network resources, capacity limited node maintains trust information for subset of nodes. The approach has been effective over Eigen approach and Peer approach (Chen, Guo and Bao, 2014). The authors have formulated adaptive security mechanisms to save critical resources since static resources consume and waste a lot of power (Hellaoui et al., 2016). It has also been concluded that mobility data of consumers have intense relation with technical, social and ethical issues (Melis et al., 2016). Quality of urban living has been further alleviated using personal data and other related information. IoT services has been evolved as Mobility as a Service (MaaS). (Yan et al., 2014) have highlighted the value of trust in IoT applications to give seamless services. Chen et al., 2011 have emphasized on the importance of CPS and IoT

Figure 4. Classification of trust models in IoT framework
(Source:https://www.researchgate.net/publication/342317404_A_DLT-based_Trust_Framework_for_IoT_Ecosystems)

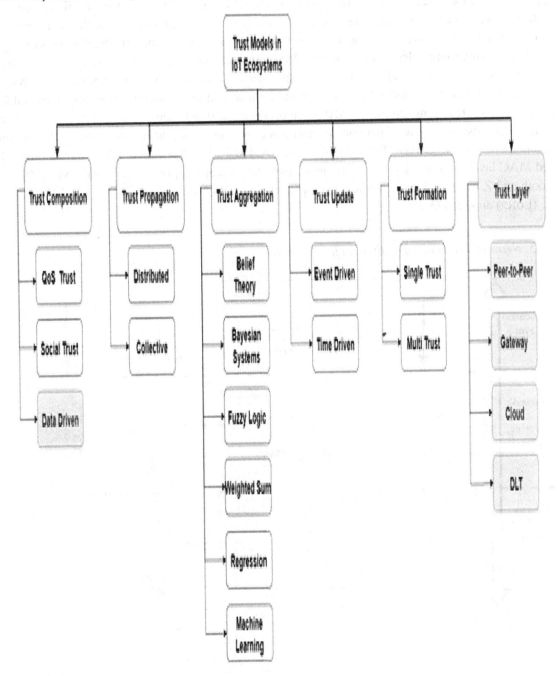

devices; since these devices are implemented at the physical layer. The authors have proposed a Trust management Model (TRM-IoT) on the basis of fuzzy logic. Simulations are done using NS-3 simulator and results yield better End to end Packet Forwarding Ratio (EPFR) at the application layer along with improved Average Energy Consumption (AEC) and better Packet Delivery Ratio (PDR). Ruan Y et al., 2018 have emphasized on trustworthiness of partners to help various agents in an IoT framework. The authors have simulated the scenario to show that analysis error is reduced in sample food scenario. Subramanian et al., 2020 have propounded that in today's world of fog computing trust is an important factor while making routing decisions. This trust factor ensures that nodes will behave cordially with in an environment. Routing protocol for low power and lossy environments (RPL) proffer very less protection in case of attacks. The authors have exemplified light weight trust analysis especially designed for fog- IoT networks. Hashemi and Aliee, 2019 have propounded that usage of single metric in standard RPL algorithm results in loss of network performance. However they have advocated Dynamic and Comprehensive Trust Model (DCTM-IoT) to propose DCTM-RPL. Oliveira et al., 2019 have thoroughly studied MAC layer protocols used for communication. Short range applications include NFC, Zigbee, Bluetooth and Z wave applications whereas long range includes Narrow band IoT (NB-IoT) and Long Range (LoRa) applications.

Figure 5. Network layers along with associated threats

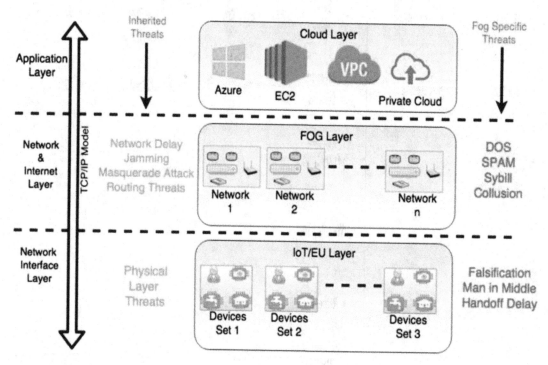

Kamble et al., 2017 have elucidated that physical devices connected to IoT networks generate enormous amount of data. The authors have classified taxonomy of attacks and suggested ways to tackle topology based attacks. Pu C., 2019 has advocated that low power lossy networks with RPL protocol are vulnerable to various Denial of Service (DoS) attacks. The author has investigated spam DIS attack and introduced novel trickle algorithm to minimize the effect of spam DIS attacks. Yavuz F et al., 2018 proposed deep learning solutions for inspecting attacks in IoT environment. The authors have used Cooja IoT simulator for detecting routing attacks on network layer in IoT framework. Different attacks at network layer such as Deceased Rank (DR), Version number modification (VN) and Hello Flood (HF) attacks have been analyzed. Different parameters like Reception Rate (RR), Reception Average Time (RAT), Received Packet Count (RCP), Transmission Rate (TR), Transmission Average Time (TAT) and Transmitted Packet Count (TCP) have been thoroughly analyzed in this paper. Importance of back propagation algorithms have been exploited in deep learning scenarios so that error gradient can be found quickly. The proposed model display 99% accuracy on the parameters F1-score and AUC score. Qin et al., 2019 have proposed Intelligent Maintenance Light Weight Anomaly Detection System (IMLADS) for IoT framework. The authors have simulated the experiments in a lab scenario with different computers, gadgets, sensors for PM measurement and temperature measurement along with wireless access points. The authors have obtained features using mobile agents of selected nodes. In order to reduce data dimensionality Principal Component Analysis (PCA) method is employed. Further, Density based Spatial Clustering Anomaly detection along with Noise (DBSCAN) is implemented to differentiate between normal and anomaly clusters. The authors have proposed the lightweight mobile agents for providing secure IoT networks. Niu et al., 2019 have advocated edge computing to handle IoT data analytics. Jamali et al., 2018 have advised a unique defense mechanism corresponding to human body against wormhole attack. Simulation results give better network performance in MANETS. Gomes et al., 2019 have highlighted that sensors have been used to collect data for more than 2300 years. The authors have advised that feedback loop between controller and plant is secured in dedicated network rather than on the internet. Ferrer et al., 2019 have analyzed MAC protocols for extending usage up to satellite IoTs. CubeSat has been implemented in satellite communication in this paper. Lamaazi and Benamar, 2019 have propounded that RPL algorithm maintains link quality using ETX metric and trickle algorithm. The authors have compared RPL-EC which is based on expected transmission count metric with RPL-FL which is based on trickle algorithm using fuzzy logic. The authors have found that network life and convergence time has fared well in RPL-EC routing algorithm. Jain and Kashyap, 2020 have emphasized that network lifetime can be prolonged by selecting nodes with higher residual energy in an ad hoc environment.

Canbalaban and Sen, 2020 have proffered a unique cross layer intrusion detection for implementing RPL algorithm for IoT devices. The authors have highlighted that lossy links in wireless networks along with resource constrained nodes made it difficult to provide a secure solution. Jain and Kashyap, 2019 have suggested Expected Transmission Count (ETX) metric for estimating link quality by incorporating both packet forwarding and packet acknowledgement probabilities. However, this technique works smoothly for homogeneous nodes. Bindel, Chaumette and Hilt, 2015 have proposed Fast ETX (F-ETX) metric for link quality estimation for Vehicular Adhoc Networks (VANETS). Sanmartin et al. have proposed the improvement in IoT framework using BF-ETX which results in reduction of network latency. Lots of routing protocols are ther for Wireless Sensor Networks (WSN) but these are limited to flat architecture. Hence there is an urgent need of another hirerachical technique which considers link quality at the network layer and trustworthiness at the service layer while making secure routing decisions for heterogeneous nodes in Service Oriented Architecture (SOA) adopted for IoT framework. Trust and Link quality ensured RPL (TL-RPL) has been proposed in the paper and performance has been compared with standard RPL and RPL-EC (Lamaazi, H. and Benamar, N., 2019).

Figure 6. Destination Oriented Directed Acylic Graph (DODAG)

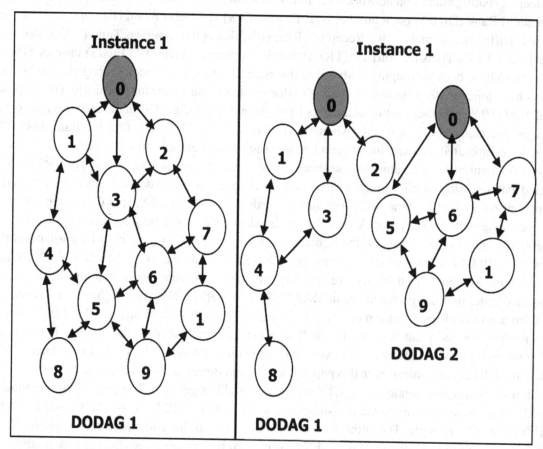

4. PROPOSED METHODOLOGY

In this paper first link quality is estimated between two interacting parties using ETX metric. Good quality links are having better reputation while poor quality link will not have good reputations. Trust is calculated among interacting parties using recommendations.

ETX of the link is calculated using eq. 1. This metric considers not only forward probability of sending packet but also reverse probability of packet acknowledgement.

$$ETX = 1/ (NLQ * LQ)\dots\dots\dots\dots\dots\dots eq. 1$$

Where LQ= Link Quality, NLQ=Neighbor Link Quality

Trust Evaluation Based on Graphs

Destination Oriented Directed Acylic Graph (DODAG) are used to represent nodes in acyclic graph as shown in Figure 6.

Table 1. Simulation parameters

Simulator Used	Cooja under Contiki OS
Number of nodes	10,20,25,30,40,50,55,60
Simulation time	400 sec
I min	210
Imax	220
Data packet	65 sec
Routing protocol	RPL, RPL-EC, TL-RPL
Interference Range	100

The nodes in the graph are labelled as trustor (source), trustee (destination) or recommender (intermediate). Edges connecting the graph nodes quantify trust in the range [0, 1]. 0 value shows no trust at all whereas value 1 depicts full trust between nodes. For example s represents source, d shows destination node and a,b are intermediate nodes. If nodes a,b in direct contact with destination node d then these nodes have idea about its trustworthiness. If nodes are not in direct contact then trustworthiness is computed using tidal trust model.

Tidal Trust

This model helps to find trusted path to generate recommendations based on certain pre-computed parameter. This algorithm explores path using quite popular shortest path Dijkstra algorithm working in Breadth First Search (BFS) fashion. Dijkstra algorithm is chosen otherwise BFS proffers latency in network by computing distance to unwanted dummy nodes. Tidal trust computes most trusted shortest path from trustor to trustee. Trust from source node to destination node is calculated by link quality metric. If link quality is above a threshold value only then that path is considered. Calculations for path computation begin from trustor node and reaches up to trustee.

PERFORMANCE METRICS

Energy Consumption

Total energy consumed by the nodes yield energy consumption. From MAC layer, it is directly proportional to the processing time comprising of transmission, reception and idle mode when communication is not direct between nodes. Energy limitations of sensor nodes is much higher since these are terminal nodes whereas base station has unlimited energy.

Packet Delivery Ratio

Ratio of successful packets received to the packets transmitted gives PDR. This metric is very important when nodes are considered from energy point of view. Retransmission may occur due to link loss properties of a wireless link which results in lower PDR.

Figure 7. Packet Delivery Ratio vs. Network Size

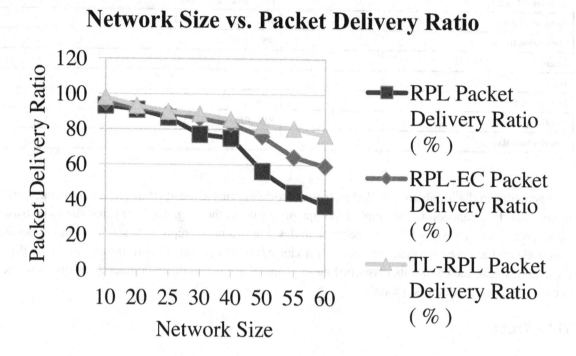

Simulation Parameters

Simulation parameters are tabulated below in Table 1.

Standard RPL provides much lower value of PDR since packet lost is higher in number due to network collision. TL-RPL performs better than RPL-EC due to the fact that as network size grows, packet collisions are lesser in number due to trust among nodes. Figure 8 displays reduced energy cost per packet in TL-RPL since lesser collisions results in reduced drop rate which further reduces energy required for retransmissions.

CONCLUSION AND FUTURE DIRECTION

In the proposed scheme network layer and service layer had been improved by ensuring better QoS and trustworthiness amongst communicating entities. Power consumption in networks using ETX metric had been reduced as more reliable links are chosen. Though QoS requirements may vary from device to device as energy minimization is very crucial parameter for battery operated devices whereas it is not that important for devices plugged with power supplies. In the future work proposed protocol should be able to handle different cyber attacks at the sensing layer. Protocol should be robust enough so that it should avoid collision among different RFID tags. Another major concern is at the interface layer since universal standards for RFID tags are still missing.

Figure 8. Network size vs. Energy cost

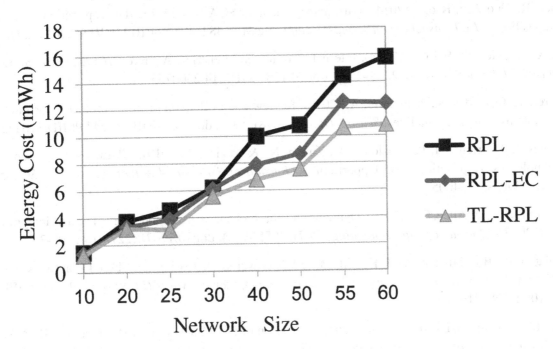

Number of nodes vs. Energy Cost per Packet

REFERENCES

Abedin, S. F., Alam, M. G. R., Kazmi, S. A., Tran, N. H., Niyato, D., & Hong, C. S. (2018). Resource allocation for ultra-reliable and enhanced mobile broadband IoT applications in fog network. *IEEE Transactions on Communications*, *67*(1), 489–502. doi:10.1109/TCOMM.2018.2870888

Asghari, P., Rahmani, A. M., & Javadi, H. H. S. (2019). Internet of Things applications: A systematic review. *Computer Networks*, *148*, 241–261. doi:10.1016/j.comnet.2018.12.008

Bindel, S., Chaumette, S., & Hilt, B. (2015, May). F-ETX: an enhancement of ETX metric for wireless mobile networks. In *International Workshop on Communication Technologies for Vehicles* (pp. 35-46). Springer. 10.1007/978-3-319-17765-6_4

Canbalaban, E., & Sen, S. (2020, October). A Cross-Layer Intrusion Detection System for RPL-Based Internet of Things. In *International Conference on Ad-Hoc Networks and Wireless* (pp. 214-227). Springer. 10.1007/978-3-030-61746-2_16

Chai, H. S., Choi, J. Y., & Jeong, J. (2015, January). An Enhanced Secure Mobility Management Scheme for Building IoT Applications. In FNC/MobiSPC (pp. 586-591). doi:10.1016/j.procs.2015.07.258

Chen, D., Chang, G., Sun, D., Li, J., Jia, J., & Wang, X. (2011). TRM-IoT: A trust management model based on fuzzy reputation for internet of things. *Computer Science and Information Systems*, *8*(4), 1207–1228. doi:10.2298/CSIS110303056C

Chen, R., Guo, J., & Bao, F. (2014). Trust management for SOA-based IoT and its application to service composition. *IEEE Transactions on Services Computing*, *9*(3), 482–495. doi:10.1109/TSC.2014.2365797

Da Xu, L., He, W., & Li, S. (2014). Internet of things in industries: A survey. *IEEE Transactions on Industrial Informatics*, *10*(4), 2233–2243. doi:10.1109/TII.2014.2300753

Ferrer, T., Céspedes, S., & Becerra, A. (2019). Review and evaluation of MAC protocols for satellite IoT systems using nanosatellites. *Sensors (Basel)*, *19*(8), 1947. doi:10.339019081947 PMID:31027250

Gomes, J., Rodrigues, J. J., Rabêlo, R. A., Kumar, N., & Kozlov, S. (2019). IoT-Enabled Gas Sensors: Technologies, Applications, and Opportunities. *Journal of Sensor and Actuator Networks*, *8*(4), 57. doi:10.3390/jsan8040057

Hashemi, S. Y., & Aliee, F. S. (2019). Dynamic and comprehensive trust model for IoT and its integration into RPL. *The Journal of Supercomputing*, *75*(7), 3555–3584. doi:10.100711227-018-2700-3

Hellaoui, H., Bouabdallah, A., & Koudil, M. (2016, November). Tas-iot: trust-based adaptive security in the iot. In *2016 IEEE 41st Conference on Local Computer Networks (LCN)* (pp. 599-602). IEEE. 10.1109/LCN.2016.101

Jain, R., & Kashyap, I. (2019). Performance Evaluation of OLSR-MD Routing Protocol for MANETS. In *Advances in Computer Communication and Computational Sciences* (pp. 101–108). Springer. doi:10.1007/978-981-13-6861-5_9

Jain, R., & Kashyap, I. (2019). An QoS aware link defined OLSR (LD-OLSR) routing protocol for MANETs. *Wireless Personal Communications*, *108*(3), 1745–1758. doi:10.100711277-019-06494-9

Jain, R., & Kashyap, I. (2020). Energy-Based Improved MPR Selection in OLSR Routing Protocol. In *Data Management, Analytics and Innovation* (pp. 583–599). Springer. doi:10.1007/978-981-32-9949-8_41

Jamali, S., Fotohi, R., & Analoui, M. (2018). An artificial immune system based method for defense against wormhole attack in mobile adhoc networks. *Tabriz Journal of Electrical Engineering*, *47*(4), 1407–1419.

Kamble, A., Malemath, V. S., & Patil, D. (2017, February). Security attacks and secure routing protocols in RPL-based Internet of Things: Survey. In *2017 International Conference on Emerging Trends & Innovation in ICT (ICEI)* (pp. 33-39). IEEE. 10.1109/ETIICT.2017.7977006

Lamaazi, H., & Benamar, N. (2019). A novel approach for RPL assessment based on the objective function and trickle optimizations. *Wireless Communications and Mobile Computing*, *2019*, 2019. doi:10.1155/2019/4605095

Li, S., Da Xu, L., & Zhao, S. (2015). The internet of things: A survey. *Information Systems Frontiers*, *17*(2), 243–259. doi:10.100710796-014-9492-7

Li, S., Da Xu, L., & Zhao, S. (2018). 5G Internet of Things: A survey. *Journal of Industrial Information Integration*, *10*, 1–9. doi:10.1016/j.jii.2018.01.005

Melis, A., Prandini, M., Sartori, L., & Callegati, F. (2016, September). Public transportation, IoT, trust and urban habits. In *International conference on internet science* (pp. 318-325). Springer. 10.1007/978-3-319-45982-0_27

Mohanta, B. K., Jena, D., Satapathy, U., & Patnaik, S. (2020). Survey on IoT Security: Challenges and Solution using Machine Learning, Artificial Intelligence and Blockchain Technology. *Internet of Things*, 100227.

Niu, Y., Zhang, J., Wang, A., & Chen, C. (2019). An efficient collision power attack on AES encryption in edge computing. *IEEE Access: Practical Innovations, Open Solutions*, *7*, 18734–18748. doi:10.1109/ACCESS.2019.2896256

Oliveira, L., Rodrigues, J. J., Kozlov, S. A., Rabêlo, R. A., & Albuquerque, V. H. C. D. (2019). MAC layer protocols for Internet of Things: A survey. *Future Internet*, *11*(1), 16. doi:10.3390/fi11010016

Pu, C. (2019, February). Spam dis attack against routing protocol in the internet of things. In *2019 International Conference on Computing, Networking and Communications (ICNC)* (pp. 73-77). IEEE. 10.1109/ICCNC.2019.8685628

Qin, T., Wang, B., Chen, R., Qin, Z., & Wang, L. (2019). IMLADS: Intelligent maintenance and light-weight anomaly detection system for internet of things. *Sensors (Basel)*, *19*(4), 958. doi:10.339019040958 PMID:30813486

Ranathunga, T., Marfievici, R., McGibney, A., & Rea, S. (2020, June). A DLT-based Trust Framework for IoT Ecosystems. In *2020 International Conference on Cyber Security and Protection of Digital Services (Cyber Security)* (pp. 1-8). IEEE.

Ruan, Y., Durresi, A., & Uslu, S. (2018, May). Trust assessment for internet of things in multi-access edge computing. In *2018 IEEE 32nd International Conference on Advanced Information Networking and Applications (AINA)* (pp. 1155-1161). IEEE. 10.1109/AINA.2018.00165

Sanmartin, P., Jabba, D., Sierra, R., & Martinez, E. (2018). Objective function BF-ETX for RPL routing protocol. *IEEE Latin America Transactions*, *16*(8), 2275–2281. doi:10.1109/TLA.2018.8528246

Subramanian, N., GB, S. M., Martin, J. P., & Chandrasekaran, K. (2020, January). HTmRPL++: A Trust-Aware RPL Routing Protocol for Fog Enabled Internet of Things. In *2020 International Conference on COMmunicationSystems & NETworkS (COMSNETS)* (pp. 1-5). IEEE. 10.1109/COMSNETS48256.2020.9027387

Ulmer, J., Belaud, J., & Le Lann, J. (2013). A pivotal-based approach for enterprise business process and IS integration. *Enterprise Information Systems*, *7*(1), 61–78. doi:10.1080/17517575.2012.700326

Whitmore, A., Agarwal, A., & Da Xu, L. (2015). The Internet of Things—A survey of topics and trends. *Information Systems Frontiers*, *17*(2), 261–274. doi:10.100710796-014-9489-2

Yan, Z., Zhang, P., & Vasilakos, A. V. (2014). A survey on trust management for Internet of Things. *Journal of Network and Computer Applications*, *42*, 120–134. doi:10.1016/j.jnca.2014.01.014

Yang, D. L., Liu, F., & Liang, Y. D. (2010, December). A survey of the internet of things. In *Proceedings of the 1st International Conference on E-Business Intelligence (ICEBI2010)*. Atlantis Press. 10.2991/icebi.2010.72

Yang, Y., Wu, L., Yin, G., Li, L., & Zhao, H. (2017). A survey on security and privacy issues in Internet-of-Things. *IEEE Internet of Things Journal, 4*(5), 1250–1258. doi:10.1109/JIOT.2017.2694844

Yavuz, F. Y., Devrim, Ü. N. A. L., & Ensar, G. Ü. L. (2018). Deep learning for detection of routing attacks in the internet of things. *International Journal of Computational Intelligence Systems, 12*(1), 39–58. doi:10.2991/ijcis.2018.25905181

Chapter 3
State of the Art Review of IIoT Communication Protocols

Shruthi H. Shetty

Sahyadri College of Engineering and Management, India

Ashwath Rao

Sahyadri College of Engineering and Management, India

Rathishchandra R. Gatti

Sahyadri College of Engineering and Management, India

ABSTRACT

Internet of things (IoT) guarantees an incredible future for the internet where the sort of correspondence is machine-machine (M2M). This arising standard of networking will impact all aspects of lives going from the computerized houses to smart IoT-based systems by implanting knowledge into the articles. This chapter intends to give an exhaustive outline of the IoT, IIoT situation and audits its empowering innovations. And finally, applications resulting from IoT/IIoT that facilitate daily needs are discussed.

1. INTRODUCTION

The Internet of Things (IoT) is expanding as a worldwide processing network where the whole world is correlated with the Internet (Farooq et al., 2015). The constant improvement of IoT applications is building an approach for progressive and inventive thoughts and arrangements, some of which are stretching the boundaries of best-in-class innovation. In the present era the idea of a "smart environment" has developed as one of the thriving technologies covering a wide scope of areas, for example biomedical, smart home, infrastructure, industrial surveillance, transport system, utilities, environmental and agronomic monitoring. The IoT and the Industrial IoT (IIoT) are developing so as to approach the up-and-coming age of Tactile IoT/IIoT, something which will unite hyperconnectivity, edge-Computing, Distributed Ledger Technologies (DLTs) and Artificial Intelligence (AI) (Eisenhauer et al., 2018). IIoT networks are the large-scale production of IoT objects for engineering applications such as smart engineering and

DOI: 10.4018/978-1-7998-7541-3.ch003

supply-chain administration. Wireless infrastructures for IIoT are an essential element of the structure that requests to satisfy the necessities such as consistent connectivity and insignificant delays. Therefore, the utilization of wireless technology is likely to modernize engineering applications and to empower the formation of an IIoT. IIoT networks usually aim on industrialized applications, such as checking and control of different devices in an industry (Jaloudi, 2019). The significance of the IoT from a modern perspective can get acknowledged once the gadgets that are associated with worldwide organization can interface with each other for the steady organization of clients, knowledge requisition for business, stock frameworks according to the customers prerequisites, and can likewise perform examination for business. Hence, enterprises are quickly embracing this innovation for imaginative innovative arrangements (Elahi et al., 2020). At present, IoT is liable for overseeing narrow systems administration framework, speaking to a somewhat enormous number of remote gadgets that can be associated to the web. With the unending flourishing of the emerging IoT progressions, the possibility of the IoT will before long be unyieldingly making enormous scope. This emerging perspective of frameworks organization will affect all parts of our lives going from the mechanized houses to sharp prosperity and atmosphere by implanting knowledge into the articles throughout.

IIot has exceptional correspondence prerequisites, as well as high consistency, low latency, workability and security. These are naturally given by, 5G versatile innovation, rendering it into an effective contender for backing up IIoT situations. 5G offers attributes fundamental for modern use-cases, for example, power, ultralow idleness, high information rates, and gigantic number of gadgets. Thus 5G is a correspondence innovation in an industrial area (Varga et al., 2020).

2. STANDARDS OF IIOT COMMUNICATION PROTOCOLS

The low-power technologies and communication protocols are developed to satisfy the rising requirements of an extensive range of IoT requests. IoT communication strategies for correspondence that ensure supreme security to the data being exchanged between IoT related gadgets. For each communication between gadgets, there is a requirement for a medium; ideally a distinctive language that all the appliance in the designated IoT environment will have the choice to comprehend. This exceptional medium is given by means of IoT protocols. The structure elements for IoT gadgets are gateways, sensors, processors, actuators and the requisition. Processors go about as the mind of IoT framework. The Passages are answerable for moving a prepared information to an appropriate area. Bluetooth, Low Power Wide Area Network (LPWAN) and WiFi are the vehicle that correspondence guidelines for the IoT. Information is gathered from the environmental factors by sensors, while actuators encompass the handled information. The gathered information must be used by an appropriate application. The fundamental utilizations of IoTs incorporate traffic the board, medical care frameworks, ecological checking, and savvy structures (Elahi et al., 2020).

Many different network/wireless protocols are used to link smart devices namely IPv6, 6LoWPAN, ZigBee, Zensys-Wave (Z Wave), Bluetooth-Low Energy (BLE), Near Field Communication (NFC) are the short-range standard net protocols, while SigFox and Cellular are LPWAN standard protocols.

SigFox

It is a type of wireless protocol which allows communication at least energy consumption. In other words, SigFox is a worldwide administrator that constructs remote organizations to interface low-control items, for example, power meters and smartwatches, which should be consistently on and emanating modest quantities of information. It utilizes the distinctive binary phase-shift keying (DBPSK) and the Gaussian frequency shift keying (GFSK) that empowers correspondence utilizing the Industrial, Scientific and Medical ISM radio band which utilizes 868MHz in Europe and 902MHz in the US. This wireless prototype can operate through a battery for a high prolonged time without the necessity of maintenance or change of battery. SigFox aims at low-power objects namely M2M application areas and sensors. It permits the transference of small quantities of data that reach up and about 50 kilometres. Prototype uses the mechanization called Ultra Narrow Band (UNB) specially outlined to handle the transfer of data at rate, 10-1000 bits per second (bps) which can run only on small batteries (Ertürk et al., 2019). It is used in infrastructure comprising consumer possessions, automotive management, smart metering etc.

Cellular

Cellular technology is preferred for applications that require long distance operation and high quantity data. It's a method of interfacing actual things (like sensors) to the web utilizing essentially a similar stuff behind a cell phone. Rather than expecting to make another, private organization to house the IoT gadgets, they can piggyback on a similar portable organization as cell phones. By taking supremacy of GSM/3G/4G networks. it provides high reliable connectivity speed/rate to the internet. Cellular networks are capable of handling huge courses of data but consume high power. Hence it is not preferable for M2M applications or for local networks. This cellular prototype is used for applications that include mobile devices and with the rise of 5G network marks cellular prototype to stand a chance. Cell IoT gives a choice to LPWAN like the non-cell LoRaWAN and Sigfox advances, which work in unlicensed groups. Cell networks fit for encouraging massive progressions of information, so there is no need of fabricating any new actual foundation to help cell IoT. In any case, for quite a while, cell empowered IoT gadgets utilized a huge load of intensity, restricting their handiness to applications where power was uninhibitedly accessible. Presently, in any case, new cell empowered sensors can move sensible measures of information across impressive distances without depleting the battery. What's more, with 5G not too far off, the future looks incredible for cell IoT.

6LoWPAN

It is the most commonly used prototype in IoT. 6LoWPAN is an IP based standard protocol with less cost and low power utilization. This standard has an opportunity of a recurrence band, actual layer and can likewise be utilized on all parts of different interchanges stages, which includes Ethernet, Wi-Fi, 802.15.4 and sub-1GHz ISM. A key property is IPv6 stack (Internet Protocol adaptation 6), which is a significant acquaintance as of late to empower the IoT. By direct route, it can be associated with other IP based grid without the help of a midway entity. It is an IP wireless prototype that uses IPv6 IEEE802.15.4 networks. 6LoWPAN suggests an adaptation layer in the mid of the MAC, the network layer which holds an interoperability amongst IEEE 802.15.4 and IPv6. Intended to consign IPv6 bundles over IEEE802.15.4 organizations and actualizing IP norms inclusive of TCP, UDP, HTTP, COAP, MQTT, and web-sockets.

The standard proffer from start-to-finish addressable hubs, permitting a switch which associates the organization to IP. Thus, 6LowPAN is a cross section grid that is self-mending, vigorous and versatile.

Wi Fi

It is the most widespread IoT protocol and a remote convention that was worked with the expectation of supplanting Ethernet utilizing remote correspondence over unlicensed groups. Wi-Fi offers fast data transmission and works optimum for LAN environments. This connection is capable of processing large amounts of information and is developed on IEEE 802.11n standard (Ruiz-Garcia et al., 2009). Wi-Fi is generally used in home and business domains which can handle transfer rates upto 100Mbps, but consume more power. Standard WiFi, while being the conspicuous decision for IoT, has impediments in both reach and energy effectiveness. The IEEE tended to these deficiencies by distributing details for 802.11ah and 802.11ax. WiFi HaLow innovation depends on the IEEE802.11ah standard was acquainted explicitly with addressing the reach and force worries of IoT. 802.11ah utilizes the 900 MHz ISM permit absolved band to furnish an all-encompassing reach with low- power necessities. The forthcoming High Efficiency Wireless (IEEE802.11ax) standard likewise adds various IoT inviting highlights. It holds the focus on wake time and station gathering highlights from 802.11ah to permit the customers to be forced frugally and evade impacts. Also, the uplink multi-client MIMO abilities, combined with the more modest (78.125 kHz) subcarrier dividing, permits up to 18 customers to send information at the same time inside a 40 MHz channel.

Zigbee

This prototype is specially designed for industrial application zones, where low power is required. It works inside the ISM band of 2.4 GHz, which permits free tasks, enormous range designation and It is built-on IEEE 802.15.4 networks. This IEEE standard characterizes the layers (physical layer and MAC) for low-velocity remote individual region organizations. The actual layer underpins three recurrence groups with various gross information rates: 2,450 MHz, 915 MHz and 868 MHz (Ruiz-Garcia et al., 2009). ZigBee is broadly used to control a few gadgets inside the scope of 10–100 m. The information pace of 250 Kbps is most appropriate for two route correspondence between a few sensors hubs and regulators. Zigbee has substantial compensations in composite systems with low power consumption, high security and is well suited for high level low cost systems with small size which require minimum data rate, long battery-life and secure networking. The most recent form of ZigBee is the as of late dispatched 3.0, basically the unification of different ZigBee remote norms into a solitary norm. Zigbee is all set to take proper advantage of remote control and sensor networks in M2M and IoT requisition.

BLE

The new BLE as it is presently marked – is a critical convention for IoT applications. Is specially designed to provide LP consumption and offer extended battery life compared to classic bluetooth. It is one of the short range communication IoT protocols, well suitable for mobile devices. Also known as smart Bluetooth specially designed for IoT applications which require low-latency, reduced power consumption and low bandwidth. Smart Bluetooth utilizes the 868 to 915 MHz and 2.4GHz band that convey at 1 Megabits for every second. At the point when Bluetooth 4.0 was first delivered, it was not focused for

IP-associated gadgets however for correspondence between two people. In any case, the most recent arrival of Bluetooth 4.2 proffer includes that makes BLE a serious up-and-comer amidst the accessible LP correspondence innovations in IoT space. Gadgets that utilize Bluetooth Smart highlights consolidate the Bluetooth Core Version 4.0 (or high up – the most recent form 4.2) with a joined essential information rate and LP center design for RF handset, baseband and convention stack. Significantly, rendition 4.2 by means of its IP Support outline will permit Bluetooth-Smart sensors to get to the web straightforwardly through 6LoWPAN availability (Raza et al., 2015).

NFC

It is a wireless prototype with short range communication where data transmission is enabled by bringing the devices in range sparsely more than inches that is NFC uses electromagnetic induction amongst two devices placed together near the arena. NFC is used not only to credentials, as well to intricate the two-way communiqué. Basically, it broadens the ability of contactless card innovation and empowers gadgets to share data a good way off that is under 4cm. It has a recurrence ISM band of 13.56 MHz on ISO/IEC 18000-3 air association and at rates going from 106 kbit/s to 424 kbit/s. NFC includes a deviser and an objective; the deviser effectively produces a RF field that can control a latent objective that empowers NFC focuses to take basic structure factors, for example, labels, stickers, key dandies, or battery-less cards. NFC distributed correspondence is conceivable given the two gadgets are fueled. NFC allows two modes of communication that is passive and active mode of communication. In passive mode, just a single NFC gadget produces a RF field. The subsequent gadget alluded to as the objective uses a procedure called load balance to move the information back to the essential gadget/initiator. whereas in active mode of communication both NFC devices produce RF fields.

Z Wave

It is a wireless model established by Zensys based on low power RF communication. This technology is highly preferable for home automation applications and small commercial fields. Z-Wave is particularly designed for small information packets at transfer rate upto 100kbps and distance of 30-meter point-to-point transmission. In this way, it is appropriate on the side of slight communications in IoT requisition, similar to light/energy control, and control of medical services. Z-Wave relies upon two kinds of gadgets (monitoring and slave). The slave hubs property is minimal effort gadgets incapable to start communication. It just answers and accomplish orders sent by administer gadgets that start communication inside the organization. Z-Wave includes a wide biological system of clever items that cooperate among label and models. With the cutting edge innovation of Z-Wave, there are no obstruction from 2.4GHz remote advancements in a comparable band. Currently, more than 60 million Z-Wave items have just been sold around the world. Shrewd homes need remote networks, and Z-wave has arisen as a definitive answer for home mechanization.

LoRaWAN

LoRaWAN is a LPWAN/Low Power-Wide Area Networking level illuminated by LoRa Alliance which possess significant attributes like integrated security and GPS- positioning. LPWAN is one of the substitute economical technologies in the IoT domain that work at lower information rates to kilometres scale inclusion of thick metropolitan to rural districts (Ertürk et al., 2019). LoRa is referred to as Long-Range and a physical-layer technology. LoRa is a RF regulation, thus compares to actual layer in the OSI reference model, while LoRaWAN is a MAC standard that arranges the medium. LoRaWAN lay hold of associations, networks, analysts and has emerged into a mainstream LPWAN innovation. This prototype is a media access control (MAC) protocol designed to wirelessly link battery functioned devices to the internet/server in local or global networks. LoRaWAN has approaches to forestall network clog permitting channels to be accessible at just %1 obligation cycle for EU 868MH. Upgraded for LP utilization and prop up huge organizations with great many gadgets, information rates scale from 0.33 kbps to 50 kbps Thus, LoRaWAN is a reduced power consumption device and supports extended range communication. At the point when, price is considered, possessing a long-life LoRaWAN prepared IoT gadgets turns into plausible arrangement. The applications are unlimited that is E- Cities/Farming, Smart Grids, atmospheric Monitoring and e-Health are well-known application territories for LPWAN advancements. These girds are good enough for open air IoT arrangements, for instance, shrewd city, air terminal, cultivating and so forth. Other than open air utilization, there is a hole in indoor arrangements that is signal propagation and variant organization issues will arise for indoor entreaty situations for various house clients (Ertürk et al., 2019).

MQTT

Message Queuing Telemetry Transport is a high-use innovation that was at first used to fabricate associations inside a satellite-based organization. The lightweight convention considered low data transfer capacity and force utilization. It is an ISO standard messaging protocol that allows bi-directional communication. It is designed as a publish/subscribe messaging protocol, that is ideal for associating far off gadgets with an insignificant organization transfer speed. MQTT is message situated. Each message is a discrete lump of information, obscure to the merchant. Each message is distributed to a location, known as a theme. Customers may buy into numerous points. Each customer bought in to a theme gets each message distributed to the point. MQTT today is utilized in a wide assortment of ventures, for example, car, producing, broadcast communications, oil and gas, and so on. MQTT is an adaptable and simple-to-utilize innovation that gives viable correspondence inside an IoT framework. For instance, IBM Watson IoT Platform utilizes MQTT as the primary correspondence convention. MQTT rather than HTTP is most appropriate for an application where transfer speed, parcel size and force are including some built-in costs. An industry generator with battery-fueled temperature and mugginess sensor can't bear to keep an association with workers each time it needs to push the deliberate qualities (occasion or message) into the cloud. MQTT is simply intended to beat such requirements where the association is kept up by utilizing a next to no power and the orders and occasions can be gotten with as meager as 2 bytes of overhead.

5G and IIoT

The fifth era of cell-based versatile correspondence design is ordinarily called 5G. It has been inspired by different components, some are absolutely identified with interchanges, for example, serving exceptionally populated territories with rapid portable access, and some are less identified with correspondences, for example, battery allotted span for more than 10 years. Traffic-affiliated inspirations incorporate an extending necessity for enhanced-mobile broadband (eMBB), super dependable and less idleness, purported basic correspondence situations and imagined massive machine type correspondence (MMTC), gigantic IoT, traffic requests. One of the new zones where cell versatile interchanges enter because of 5G is modern IoT, particularly with respect to super dependable and low-inertness correspondence needs (Varga et al., 2020). The critical capacities of 5G go about as an empowering influence for Industry 4.0. Industry 4.0 is characterized as the latest thing of mechanization and information trade in producing advances. Its primary exploration areas incorporate cyber physical frameworks, distributed computing and intellectual figuring. The 5G innovation is featured as it addresses the significant difficulties of a cell network like Enormous bandwidth. Higher information rate, Enormous network, low start to finish latency, financially effective Gadget computational capacities, that all are adequately contrasted with its archetypes (Shafique et al., 2020). The first advantages in 5G are the accessibility of a rich range by using the bountiful mmWave band which isn't utilized in 1G/2G/3G/4G remote correspondence framework. At mmWave, there is great path-loss and higher affectability to the climate than at cmWave. The mmWave range is along these lines broadly perceived to be utilized for short distance interchanges. Certain changes of the structure of organizations must be made to utilize mmWave at its maximum capacity (Lu et al., 2018).

Industry 4.0 imprints the move from inheritance frameworks to associated advancements, introducing keen production lines of things to come. Utilizing IoT-empowered associated gadgets, sensors, edge registering, self-mending organizations, advanced mechanics and robotization will help these modern manufacturing plants make more educated, decentralized choices, improving in general hardware and cycle proficiency. In any case, network is the bedrock of Industry 4.0. Modern organizations will require a steady, secure and quick association with catch and cycle information continuously for plant and gear observing and support. The 5G is required to give a last-mile network by giving the speed, dependability, limit and portability that makers need for fruitful IoT usage.

3. STATE OF THE ART IMPLEMENTATION OF IIOT COMMUNICATION PROTOCOLS

Agriculture

Smart Farming: IoT should be broadly tested in order to get generally applied in different agrarian applications. The Smart Farming system uses solar energy to monitor the soil using sensors. The IoT sensor hubs which are fuelled by sunlight-based energy are utilized for checking and control of horticulture fields. The observing and control in agribusiness fields incorporate activities like yields the executives and reaping, water supply control, creature control, pesticide circulation, moistness and temperature estimation applications. This prototype uses sensors which track the weather change and soil condition and verify the analysed data with the plant database to provide the customise guidance. This detected information is fed to Wi-Fi model, later the deliberate information is then uploaded to the Cloud. From this IoT Cloud the information is fed to GSM or router. The client here notices the deliberate information with the help of a control entreaty built on the client's cell phone or System (Sharma et al., 2019).

Healthcare

IoT based Healthcare: IoT empowered gadgets have made distant observing in the medical care area conceivable, releasing the possibility to keep patients protected and engaging doctors to convey standout care. It has likewise expanded patient commitment and fulfilment as communications with specialists have gotten simpler and more productive. Medical care IoT gadgets can put in energy harvesting (EH) strategy that utilize the energy from surroundings, for example, the encompassing RF to expand the battery-life. Medical care IoT- EH devices need to oppose snoops that examine the detecting information through radio channels to uncover the client area and propensities, for example, the utilization design security. This healthcare system utilizes various sensors to quantify and assess the medical care information, for example, the circulatory strain and electrocardiograms give the crisis-care and tele-health counsel. With both power and EH unit, the IoT gadget can locally measure some calculation errands (Min et al., 2019).

Implantable clinical gadgets: In the previous many years, implantable clinical gadgets have been planned and executed to notice human actual activities, improve the usefulness of some harmed or de-based organs, and convey drugs for the treatment of uncommon illnesses. Different clinical gadgets, for example, heart pacemakers, cochlear inserts, tissue triggers, etc have been generally used to give actual treatment just as help medical care following administrations in clinical practices (Ma et al., 2020).

Logistics

The rise of the IIoT advanced new challenges in the logistic area, which may require innovative changes, for example, high requirement for transparency (store network perceivability); controlled integrity (right items, at the perfect time, place, amount condition and at the correct expense) in the stock chains. These evolvements present the idea of Logistics 4.0 (Barreto et al., 2017). Supply chain, vehicle monitoring, stock administration, safe transportation and robotization of cycles are the way to IoT applications and the chief segments of associated coordination's frameworks.

Smart location management: In the logistics area, IoT can make a brilliant area the executives framework that will empower organizations to effortlessly follow driver exercises, vehicle area, and conveyance status. Whenever merchandise is conveyed or gone to a specific spot, a supervisor is told by a push message. Such an answer is an indispensable right hand in conveyance arranging and gathering and survey of timetables. All progressions are immediately recognized and reflected continuously. Along these lines, IoT innovation can be effectively utilized for improving the area of executives and smoothing out business measures.

Infrastructure

Fleet Tracing: The patented Fleet Tracking uses kinetic energy source and uses piezoelectric energy harvester (EH) technology (Shirvanimoghaddam et al., 2019). This electromagnetic EH converts mechanical energy produced by vibration/kinetic energy to electrical energy, which thus controls the Wireless Sensor Nodes. These hubs send on-going information back to the work area or cell phone of the resource proprietor. The energy gatherer is intended to last more than 100 years without support, and the sensor hubs 20 years, effectively outliving other battery-just powered frameworks. This demonstrated, safeguard innovation is additionally quick and simple to introduce. The vibration EH and wireless sensor hubs can be utilized to screen significant hardware and resources over a wide assortment of ventures, yet most regularly utilized in rail.

Smart Street Lights: Smart Solar/Kinetic Street Lights are perfect to be implemented at pedestrian regions where individuals can be directly involved in generating energy. Every single footstep generates 4 to 8 watts. Smart lights use kinetic/ pressure energy as an energy source.

Foot pavement: the footsteps are converted into electricity, which are later stowed in a battery or straight fed to devices. A distinctive tile is made of reused polymer, with the topmost surface produced using reused truck tires. A foot step that pushes down a solitary tile by five millimetres produces somewhere in the range of one and seven watts. These tiles create power/electricity with a hybrid arrangement of systems that incorporate the piezoelectric impact and acceptance, which uses copper loops and magnets.

Transportation

Transportation is a significant piece of a general public along these lines all the connected issues should be appropriately tended to. There is a requirement for a framework that can improve the transport circumstance

Smart Traffic: For a wise traffic observing framework, acknowledgment of a legitimate framework for programmed ID of vehicles and other traffic factors is significant for which we need IoT advances as opposed to utilizing regular picture handling techniques. The insightful traffic observing framework will give a decent transportation experience by facilitating the blockage. It will give highlights like burglary discovery, detailing of auto collisions, less ecological contamination. The streets of this savvy city will give redirections with climatic changes or startling gridlocks due to which driving and strolling courses will be advanced. The traffic lighting framework will be climate versatile to save energy. Accessibility of parking spots all through the city will be available by everybody (Farooq et al., 2015).

Self-Driving vehicles: these are the vehicles that are equipped for detecting the general climate through a scope of sensors and association frameworks and to decipher this data to recognize the most reasonable courses, maintain a strategic distance from hindrances and securely move with restricted or no human info.

Environment

Shrewd Environment: Forecast of catastrophic events, for example, flood, fire, seismic tremors and so on will be conceivable because of creative advances of IoT. There will be a legitimate observation of air contamination in the climate.

Smart House: IoT will likewise give answers for Home Automation with which we will have the option to distantly control our apparatuses according to our requirements. Appropriate observation of utility meters, energy and water supply will help saving assets and recognizing sudden over-burdening, water spills and so forth There will be legitimate infringement identification framework which will forestall thefts. Planting sensors will have the option to quantify the light, dampness, temperature, dampness and other cultivating vitals, just as it will water the plants as indicated by their requirements (Farooq et al., 2015).

Pipeline/Industry Monitoring: continuous checking is finished by the estimation of stream rate (pace of stream of the fluid) and it very well may be accomplished by utilizing a stream sensor. The Pipeline Surveillance System is enormously significant, that tracks the immense measure of wastage of water brought about by spillages and other conceivable pressure driven failures. Likewise, to plan a precise

Water System the executives speaks to a basic undertaking that forces a genuine report and a satisfactory arrangement particularly in the mechanical area. IoT innovation dependent on astute sensors and actual items is actualized to oversee the Aqua Distribution arrangement and to adapt to fluid wastage throughout the stockpile cycle. Here, an incredible scope of IIoT strategies have been, as of late, created and conveyed (Abdelhafidh et al., 2017).

Manufacturing

The cutting edge fabrication industry is putting resources into new advancements, for example, the IoT, huge information examination, distributed computing and network safety to adapt to framework unpredictability, increment data perceivability, improve creation execution, and gain upper hands in the worldwide market. These advances are quickly empowering another age of keen assembling (Yang et al., 2019). Alongside improving the adequacy of assembling activities, the IoT is applied in assembling to guarantee appropriate resource use, broaden hardware administration life, improve dependability, and give the best profit for resources. The IoT applications encouraging modern resource the executives include: Mechanical resource following, Stock administration, Prescient upkeep.

Smart resource monitoring: arrangements dependent on RFID and IoT are required to overwhelm conventional, accounting page based techniques by giving exact ongoing information about big business' resources, their statuses, areas and developments, IoT-based resource the executives arrangements eliminate the following weight from the workers and dispense with blunders bound to the manual strategies for information input.

IoT-based cloud producing: IoT powers expanding interests to plan and grow new framework foundations that coordinate WSNs and distributed computing into assembling settings. For instance CCIoT-CMfg, in this architecture the four layer framework gives an occasion to cloud-based assembling administration age, the executives and applications (Yang et al., 2019). cloud fabricating (CMfg) as another help situated assembling mode has given a vast consideration over the globe. Nonetheless, the critical advances on the side of actualizing CMfg is the manner by which to acknowledge producing asset astute discernment and access (Tao et al., 2014).

REFERENCES

Abdelhafidh, M., Fourati, M., Fourati, L. C., & Abidi, A. (2017). Remote Water Pipeline Monitoring System IoT-Based Architecture for New Industrial Era 4.0. *2017 IEEE/ACS 14th International Conference on Computer Systems and Applications (AICCSA)*, 1184-1191. 10.1109/AICCSA.2017.158

Barreto, L., Amaral, A., & Pereira, T. (2017). Industry 4.0 implications in logistics: An overview. *Procedia Manufacturing, 13*, 1245-1252. doi:10.1016/j.promfg.2017.09.045

Eisenhauer, M., Vermesan, O., Serrano, M., Guillemin, P., Sundmaeker, H., Tragos, E., Valiño, J., Copigneaux, B., Presser, M., Aagaard, A., Bahr, R., & Darmois, E. (2018). *The Next Generation Internet of Things – Hyperconnectivity and Embedded Intelligence at the Edge*. Academic Press.

Elahi, H., Munir, K., Eugeni, M., Atek, S., & Gaudenzi, P. (2020). Energy Harvesting towards Self-Powered IoT Devices. *Energies, 13*(21), 5528. doi:10.3390/en13215528

Ertürk, M. A., Aydın, M. A., Büyükakkaşlar, M. T., & Evirgen, H. (2019). A Survey on LoRaWAN Architecture, Protocol and Technologies. *Future Internet, 11*(10), 216. doi:10.3390/fi11100216

Farooq, M., Waseem, M., Mazhar, S., Khairi, A., & Kamal, T. (2015). A Review on Internet of Things (IoT). *International Journal of Computers and Applications, 113*, 1–7. doi:10.5120/19787-1571

Iannacci, J. (2018). Internet of things (IoT); internet of everything (IoE); tactile internet; 5G – A (not so evanescent) unifying vision empowered by EH-MEMS (energy harvesting MEMS) and RF-MEMS (radio frequency MEMS). *Sensors and Actuators. A, Physical, 272*, 187–198. Advance online publication. doi:10.1016/j.sna.2018.01.038

IoT Central. (n.d.). https://www.iotcentral.io/blog/iiot-protocols-for-the-beginners

Jaloudi, S. (2019). Communication Protocols of an Industrial Internet of Things Environment: A Comparative Study. *Future Internet., 11*(3), 66. Advance online publication. doi:10.3390/fi11030066

Johan, J. (2018). Smart Soil Parameters Estimation System Using an Autonomous Wireless Sensor Network with Dynamic Power Management Strategy. *IEEE Sensors Journal, Volume, 18*(21), 8913–8923. doi:10.1109/JSEN.2018.2867432

La Rosa, R., Livreri, P., Trigona, C., Di Donato, L., & Sorbello, G. (2019). Strategies and Techniques for Powering Wireless Sensor Nodes through Energy Harvesting and Wireless Power Transfer. *Sensors (Basel), 19*(12), 2660. doi:10.339019122660 PMID:31212839

Lu, Y., Richter, P., & Lohan, E. S. (2018). *Opportunities and Challenges in the Industrial Internet of Things based on 5G Positioning.* . doi:10.1109/ICL-GNSS.2018.8440903

Ma, D., Lan, G., Hassan, M., Hu, W., & Das, S. K. (2020). Sensing, Computing, and Communications for Energy Harvesting IoTs: A Survey. IEEE Communications Surveys & Tutorials, 22(2), 1222-1250. doi:10.1109/COMST.2019.2962526

Min, M., Wan, X., Xiao, L., Chen, Y., Xia, M., Wu, D., & Dai, H. (2019, June). Learning-Based Privacy-Aware Offloading for Healthcare IoT With Energy Harvesting. *IEEE Internet of Things Journal, 6*(3), 4307–4316. doi:10.1109/JIOT.2018.2875926

Patil, K. A., & Kale, N. R. (2016). A model for smart agriculture using IoT. *IEEE International Conference on Global Trends in Signal Processing, Information Computing and Communication*, 543-545. 10.1109/ICGTSPICC.2016.7955360

Raza, S., Misra, P., He, Z., & Voigt, T. (2015). Bluetooth smart: An enabling technology for the Internet of Things. *2015 IEEE 11th International Conference on Wireless and Mobile Computing, Networking and Communications (WiMob)*, 155-162. 10.1109/WiMOB.2015.7347955

Ruiz-Garcia, L., Lunadei, L., Barreiro, P., & Robla, J. I. (2009). A review of wireless sensor technologies and applications in agriculture and food industry: State of the art and current trends. *Sensors (Basel), 9*(6), 4728–4750. doi:10.339090604728 PMID:22408551

Shafique, K., Khawaja, B., Sabir, F., Qazi, S., & Mustaqim, M. (2020). Internet of Things (IoT) For Next-Generation Smart Systems: A Review of Current Challenges. In *Future Trends and Prospects for Emerging 5G-IoT Scenarios.* IEEE Access. . doi:10.1109/ACCESS.2020.2970118

Sharma, H., Haque, A., & Jaffery, Z. (2019). *Smart Agriculture Monitoring using Energy Harvesting Internet of Things*. EH-IoT.

Shirvanimoghaddam, M., Shirvanimoghaddam, K., Abolhasani, M. M., Farhangi, M., Zahiri Barsari, V., Liu, H., Dohler, M., & Naebe, M. (2019). Towards a Green and Self-Powered Internet of Things Using Piezoelectric Energy Harvesting. *IEEE Access: Practical Innovations, Open Solutions, 7,* 94533–94556. doi:10.1109/ACCESS.2019.2928523

Tao, F., Zuo, Y., Xu, L. D., & Zhang, L. (2014, May). IoT-Based Intelligent Perception and Access of Manufacturing Resource Toward Cloud Manufacturing. *IEEE Transactions on Industrial Informatics, 10*(2), 1547–1557. doi:10.1109/TII.2014.2306397

Varga, P., Peto, J., Frankó, A., Balla, D., Haja, D., Janky, F., Soós, G., Ficzere, D., Maliosz, M., & Toka, L. (2020). 5G support for Industrial IoT Applications— Challenges, Solutions, and Research gaps. *Sensors (Basel), 20*(3), 20. doi:10.339020030828 PMID:32033076

Yang, H., Kumara, S., Bukkapatnam, S., & Tsung, F. (2019). The Internet of Things for Smart Manufacturing: A Review. *IIE Transactions, 51*(11), 1–35. doi:10.1080/24725854.2018.1555383

Chapter 4
Energy Optimization in a WSN for IoT Applications

Syed Ariz Manzar
Amity University, India

Sindhu Hak Gupta
Amity University, India

Bhavya Alankar
Jamia Hamdard, India

ABSTRACT

Energy consumption has become a prime concern in designing wireless sensor networks (WSN) for the internet of things (IoT) applications. Smart cities worldwide are executing exercises to progress greener and safer urban situations with cleaner air and water, better adaptability, and capable open organizations. These exercises are maintained by progresses like IoT and colossal information examination that structure the base for smart city model. The energy required for successfully transmitting a packet from one node to another must be optimized so that the average energy gets reduced for successful transmission over a channel. This chapter has been devised to optimize the energy required for transmitting a packet successfully between two communicating sensor nodes using particle swarm optimization (PSO). In this chapter, the average energy for successfully transmitting a packet from one node to another has been optimized to achieve the optimal energy value for efficient communication over a channel. The power received by the sensor node has also been optimized.

INTRODUCTION

Internet of things (IoT), a phenomenon that connects various things or objects also referred as internet of nonliving things, these non-living things can be made communicable with the help of sensor nodes. WSN is mainly an innovation utilized inside an IoT framework, a large number of sensors in a WSN can be utilized to independently assemble information and send information through a switch to the internet

DOI: 10.4018/978-1-7998-7541-3.ch004

in an IoT framework. IoT is building up importance in wireless communication so that the things can be controlled and monitored wirelessly. The IoT is permitting things to stay associated and convey whenever at wherever with anything and anybody utilizing any way, any system or any service. In an IoT system the connecting devices need to be smart by making them intelligent using technologies like WSN, radio frequency identification (RFID), mobile communication and internet. Various other technologies like Machine to Machine (M2M), Vehicular to Vehicular communication (V2V) can be implemented by this IoT (Al-Garadi et al., 2020).

The mobile communication technology and internet plays an important role for setting up a communication channel for an IoT system, thus IoT is a network of things (physical) controlled by the internet. The IoT devices such as sensors, lights, meters etc. are used by smart cities to gather and break down data, the urban areas make use of this information to enhance foundation, open utilities, administrations and other services.IoT-enabled smart city arrangements help to streamline squander gathering plans by following waste levels, just as giving course improvement and operational examination. Each waste compartment gets a sensor that assembles the information about the degree of the loss in a holder (Li, 2019).

The WSN acts as the foundation for IoT applications, remote sensor frameworks have dynamically become promoters of colossal information. The ongoing sending of remote sensor arranges in Smart City establishments has incited a great deal of data being made each day over a collection of zones, with applications incorporating systematic checking, human administration observing and vehicular checking.As shown in Fig.1the fundamental components of the Smart City are smart hospitals, smart environment, keen vitality, smart security, smart office, smart private structures, smart organization and smart vehicles.Energy is a fundamental segment of life in the urban areas, as it bolsters the entire range of their financial exercises and ties down a specific degree of personal satisfaction to occupants. To meet open approach destinations under these conditions, smart cities need to change and create in a shrewd manner, without dismissing the issues of energy efficiency and supportability. The current work has been carried to optimize the energy in a WSN for IoT applications.

Figure 1. Architecture of a smart city

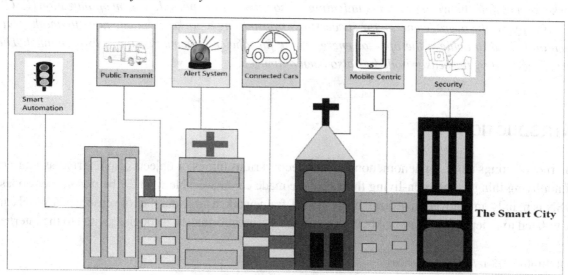

WSN, an emerging technology characterized by limited power consumption and energy storage, depends on power gathering systems requires a profound and cautious demonstration because of constrained energy supply (Alghamdi, 2020). WSN is one of the basic building blocks of IoT, consisting of a finite number of sensing nodes and these nodes communicate in a wireless multi-hopped fashion (Chaturvedi & Shrivastava, 2020) and the operation of these sensing nodes has emerged as an active research area. A number of complexities are experienced in designing a WSN for a particular application like the number of sensing nodes required, the energy consumption, lifetime of sensing nodes, type of information to be sensed, and location of sensing nodes. The beneficial use of WSN's is particularly in far-flung areas or inaccessible places and in these places WSN's faces the major challenge of limited power (Liang et al., 2019).

Due to limited capacity of the batteries driving these sensing nodes the lifetime limitation of the WSNs has become a primary concern. It would be impossible to change the energy source of these sensing nodes at the inaccessible places, therefore there is a need for energy modelling and energy management schemes that will consider all the power consuming sources and various scenarios that occur during sensing and communication operation (Abo-Zahhad et al., 2015). Increasing the lifetime of sensing nodes is one of the important concerns while dealing with wireless sensor networks. A major portion of the energy is consumed in radio transmission, the sending and receiving processes need more energy than processing the data. ZigBee is one such technology that enables WSN to operate with minimum consumption of energy so that its lifetime gets increased (Pinto et al., 2012). As the nodes are driven by battery therefore energy conservation is important and most of the energy is consumed during transmission and reception processes, the Medium access control (MAC) protocols can deal with this consumption of energy (Saha et al., 2011; Shafiabadi et al., 2019).

In this paper, we will consider some parameters like physical layer parameters and MAC layer parameters and depending on these parameters we will present an energy model and will use this model for optimizing the energy consumed for transferring a bit successfully over a channel without error.

2 ENERGY MODEL

WSN consists of a large number of nodes deployed at different locations separated by some distance (d1, d2, d3, d4 .....) as shown in Fig.2. These nodes in WSN are battery driven with limited energy

Figure 2. Single-hop communication & Multi-hop communication

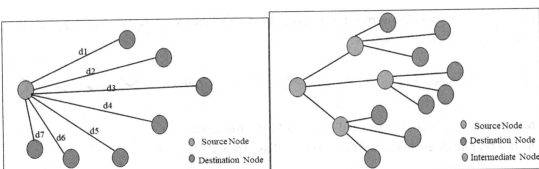

which in turn limits the energy for sending the data by the nodes (Amruta & Satish, 2013), therefore the data reaches the sink node sent by source node through multiple hops (Papageorgiou, 2003). In a WSN the communication can either be a single hop communication or multi hop communication, in a single hop communication the data packets leave the source node and take a single hop before reaching the destination node as shown on the left side of Figure 2, while as in a multi hop communication the data packets leaves the source node and take two or more hops before reaching the destination node as given in the right side of Figure 2.

Our intent is reduction in consumption of energy in nodes, therefore improving link quality in WSN and increasing the lifetime of the network. In the end we will determine the energy required for transferring the data bit successfully without any error.

2.1 Structure of Packet

The general structure of a packet in a communication system is shown in Figure 3 (Abo-Zahhad et al., 2015).

Figure 3. Structure of packet at MAC and Physical Layer

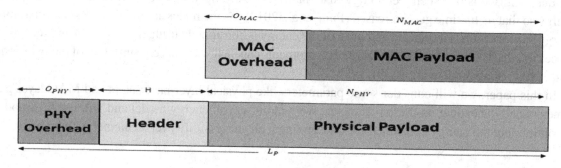

In a packet the number of bits, L_P has been evaluated in (Abo-Zahhad et al., 2015) and is given by (1);

$$L_P = O_{PHY} + H + N_{PHY} \quad (1)$$

Where, O_{PHY} is the physical overhead, H is the bits in header and N_{PHY} is the bits in physical payload represented by (2) [6] respectively.

$$N_{PHY} = O_{MAC} + N_{MAC} \quad (2)$$

Where N_{MAC} is MAC data payload and O_{MAC} is MAC overhead.

The raw bit rate R_r evaluated in [6] is expressed as given in (3);

Figure 4. Listen and sleep periods for S-MAC

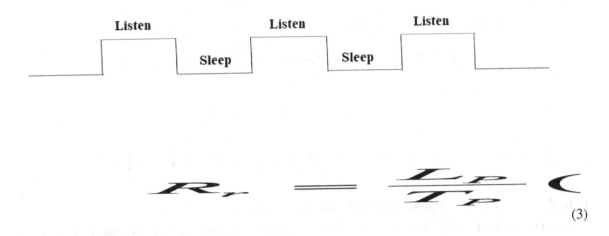

$$R_r = \frac{L_P}{T_P} \qquad (3)$$

Here T_P is the time taken to send a packet of size L_P bits and is given by (4) [6],

$$T_P = \frac{N_{MAC}}{R_A} \qquad (4)$$

Here R_A is the rate of bits for MAC payload.

2.2 Energy Consumption

To limit the energy utilization somewhat, in the current work we do optimization from different sides. The energy consumed per successfully transferred bit is obtained using (5) as in (Abo-Zahhad et al., 2015).

$$E_b = \frac{P * T_{DATA}}{N_{MAC}\left(1 - Xe^{-\frac{Y}{2BN_0 A_0 d}P_t}\right)} \qquad (5)$$

X and Y are modulation scheme dependent parameters. In this paper we have used a modulation scheme called Binary Phase Shift Keying (BPSK) for which X = 0.5 and Y = 1.

The most widely used MAC protocol in WSNs for energy conservation is S-MAC, S-MAC has predefined and constant listen and sleep period therefore having a fixed duty cycle as shown in Fig 4 (Shafiabadi et al., 2019). The substitute to original SMAC is dynamic SMAC (DS-MAC) protocol, DS-MAC has a variable duty cycle and is shown in Fig.5 (Shafiabadi et al., 2019). This duty cycle can be increased by increasing the frequency of listen and sleep period.

In a S-MAC protocol the communication between sensing nodes starts by exchanging the packets by carrier sensing (CS) for collision avoidance, after this a request to send (RTS) and clear to send (CTS) is sent. For a S-MAC protocol the power consumption is given in average and is represented as average power given using (6) (Abo-Zahhad et al., 2015);

Figure 5. Listen and Sleep periods for DS-MAC

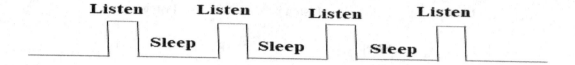

$$P = t_{tx}P_{tx} + t_{rx}P_{rx} + (1 - t_{tx} - t_{rx})P_{slp} \quad (6)$$

Here P_{slp} is the amount of power wasted during sleep state.

For a transmitting node the timing activities for transmission and reception are presented in the normalized form and are given by (7) and (8) respectively (Abo-Zahhad et al., 2015).

$$t_{tx} = \frac{\left(3t_{st} + \dfrac{L_{sync}}{R_r} + \dfrac{L_{RTS}}{R_r} + \dfrac{L_{Data}}{R_r}\right)}{T_{Data}} \quad (7)$$

$$t_{rx} = \frac{\left(t_{st} + \dfrac{L_{CTS}}{R_r}\right)}{T_{Data}} \quad (8)$$

The sensor nodes forwards their output to next nodes after measuring the incoming sensor sample and the time required for sensing and forwarding it is called data generation interval (T_{Data}).The value of t_{st} for current work is given Table 1. L_{CTS} is the size of CTS packet, L_{RTS} is the size of RTS, L_{Data} is the size of Data packet, and L_{sync} is the size of Sync packet.

In a transceiver the energy on transmitter end is consumed in the major components like digital baseband circuits, analog circuits, and power amplifiers. The total power consumed on transmitter end is given in (9) (Abo-Zahhad et al., 2015).

$$P_{tx} = P_{Dct} + P_{Act} + P_{amp} + P_t = P_{to} + (1 + \emptyset)P_t \quad (9)$$

Where, P_{Dct} is the consumption of power in digital circuitry, P_{Act} is the consumption of power in analog circuitry and P_{amp} is consumption of power in power amplifier on the transmitting end, P_t is the transmission power and Ø is a parameter that depends on drain efficiency and peak to average ratio and for current work the value of P_{to} and Ø is given in Table 1.

On the receiving end the power consumption is given using (10) (Abo-Zahhad et al., 2015);

$$P_{rx} = P_{Dcr} + P_{Acr} + P_{LNA} = P_{ro} \tag{10}$$

Here P_{Dcr} is power consumption in digital circuitry, P_{Acr} is power consumption in analog circuitry and P_{LNA} is power consumption in low noise box containing amplifier (usually LNA) (Abo-Zahhad et al., 2015), and P_{ro} for current work is given in Table 1.

For simulation purpose the various parameters in equations 5, 6,7,8,9 and 10 are given in Table 1 (Abo-Zahhad et al., 2015).

2.3 Transmission Power

The transmission power (P_t) and its relation to received power (Pr) can be given by (11) (Abo-Zahhad et al., 2015);

$$P_t = A_o d^{\alpha} P_r \tag{11}$$

Here A_o is parameter that relies upon transmitter and recipient gains and on the frequency of transmitted signal. The d in meters is the distance between transmitting and receiving ends. The parameter α is the coefficient of path loss.

P_r is received power and is given by (12) (Abo-Zahhad et al., 2015):

Table 1. Simulation parameters (Abo-Zahhad et al., 2015)

Parameter	Value	Parameter	Value
T_{DATA}	1 sec	P_{ro}	22.2 mW
L_{RTS}	30 Bytes	d	50 M
L_{sync}	24 Bytes	P_{to}	15.9 mW
L_{CTS}	24 Bytes	P_{slp}	37 μW
L_{Data}	82 Bytes	R_r	20 kbps
t_{st}	195 μsec	H	3 Bytes
A_O	40 dB	α	3.2
B	100 KHz	N_{MAC}	62 Bytes
$\emptyset$	1.86	N_{PHY}	72 Bytes
O_{MAC}	10 Bytes	----	----

$$P_r = 2\gamma B N_o \qquad (12)$$

Here γ is the Signal to Noise ratio (SNR) per bit. B, the bandwidth of signal passing through channel affected by additive white Gaussian noise (AWGN) having power spectral density N_o.

The values of the above parameters considered for simulation in the current work are given in the Table 1 (Abo-Zahhad et al., 2015).

3 OPTIMIZATION

3.1 Formulation: On rearranging (5) and (6), the energy consumed per successfully transferred bit can be written as given in (13). The objective function for the current work is given in (13).

$$E_b = \frac{[t_{tx}P_{tx} + t_{rx}P_{rx} + (1 - t_{tx} - t_{rx})P_{slp}] * T_{DATA}}{N_{MAC}\left(1 - Xe^{-\frac{\gamma}{2BN_0A_0d}P_t}\right)} \qquad (13)$$

E_b = energy consumed per successfully transferred bit.

t_{tx} = timing activity for transmission given by (8).

t_{rx} = timing activity for reception given by (9).

P_{tx} = total power consumed on transmitter end given by (10).

P_{rx} = total power consumed on receiving end given by (11).

T_{DATA} = data generation interval, for simulation basis the value is shown in Table 1.

B = bandwidth of signal, for simulation basis the value is shown in Table 1.

N_O = power spectral density of channel affected by additive white Gaussian noise, for simulation basis the value is shown in Table 1.

A_O = parameter depending on transmitter and receiver gains, for simulation basis the value is shown in Table 1.

d = distance between sensing nodes.

Table 2. Parameters for PSO

Parameter	Value
Population Size	40
Final Weight,w1	0.4
Initial Weight,w2	0.9
Number of Iterations, $I_t max$	200
Acceleration Constant (C_1, C_2)	1.4455,1.4455

P_t = transmit power.

P_{slp} = power wasted during sleep state, for simulation basis the value is shown in Table 1.

N_{MAC} = MAC data payload, for simulation basis the value is shown in Table 1.

X = 0.5 and Y = 1 for BPSK modulation scheme considered for the current work.

3.2 Optimizing the Energy Required to Transmit a bit Successfully

From (13) it can be concluded that E_b is dependent on t_{tx}, t_{rx}, P_{tx}, P_{rx}, T_{DATA}, N_O, A_O, P_t, P_{slp}, N_{MAC}, d, X, Y, B. Out of these parameter P_t and d are free parameters. With the help of PSO optimization will be attained, where P_t and d will be optimized at the same time which in turn will optimize the E_b.

PSO is an algorithm developed by Kennedy and Eberhart. It is an optimization technique based on swarm, the authors were propelled by the social conduct of creatures like birds, fish, herds and ants. PSO has been used in various areas of optimization, this algorithm searches for optimal solution and has gained importance for solving complex and hard optimization problems (Koohi & Groza, 2014). In PSO a swarm of particles move in N dimensional space to reach the optimal solution (Gupta et al., 2016). The particles in this swarm have some important properties like the current position of the particle and the velocity of particle, these particles alter their situation so that an ideal or near ideal arrangement is reached. The parameters of PSO are given in Table 2 (Gupta et al., 2016) and flow chart of PSO is given in Fig.7 (Liu et al., 2019).

The problem arises because the behaviour of P_t and d is contrary to each other, the rise in one value leads to the drop in E_b, while the rise in other value rises the E_b. To get optimal value of P_t and d at which E_b will be least is the requirement, PSO algorithm has been used to get these optimal values. By using the optimal value of P_t = 0.00031 Watts and d = 44.5 meters, theoptimized value of E_b has been attained and thus PSO has been used for attaining the optimization of E_b.

Figure 6. Article swarm optimization algorithm (Flow graph)

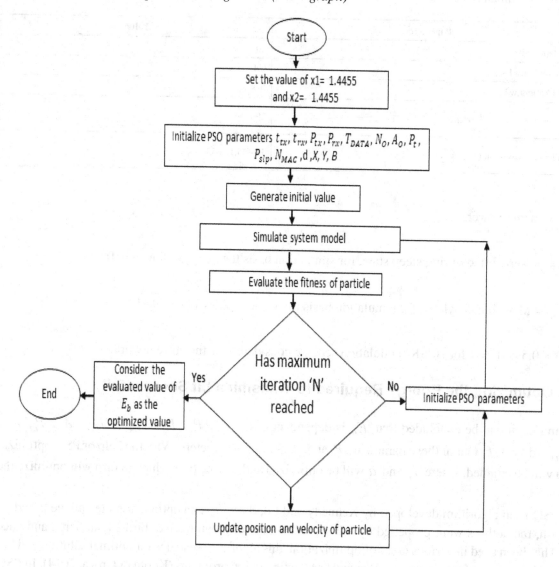

4 RESULTS AND DISCUSSIONS

The simulation has been carried out for wireless sensor networks for IoT applications regarding transmission power to contrast the performance of the network with the network given in (Abo-Zahhad et al., 2015). Also energy required to transmit a bit successfully has been optimized by using PSO for the considered scenario. The plot in Fig.8 shows the optimized energy ($E_b 2$) and non-optimized energy ($E_b 1$) clearly indicates optimized energy ($E_b 2$) is less than non-optimized energy ($E_b 1$). With this decrease in energy the overall average efficiency of the network is increased and on optimizing the energy using PSO in the current work the average efficiency of 6.83 percent is attained. The plot in Fig.9 shows optimized and non-optimized energy when plotted against the distance between the sensing nodes

Figure 7. Non-optimized average energy consumed per bit (E_b1) and optimized average energy consumed per bit (E_b2) varying with parameter transmit power (P_t)

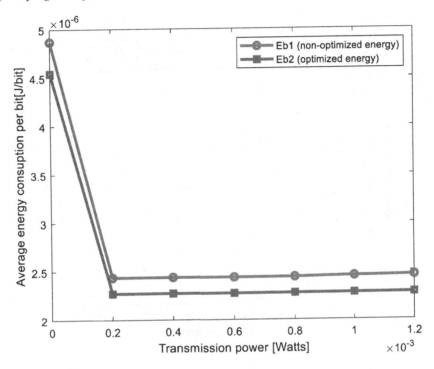

Figure 8. Non-optimized average energy consumed per bit (E_b1) and optimized average energy consumed per bit (E_b2) varying with parameter distance (d)

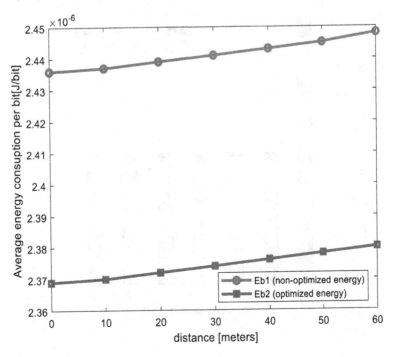

Figure 9. Non-optimized Received power ($P_r 1$) and optimized Power ($P_r 2$) varying with parameter distance (d)

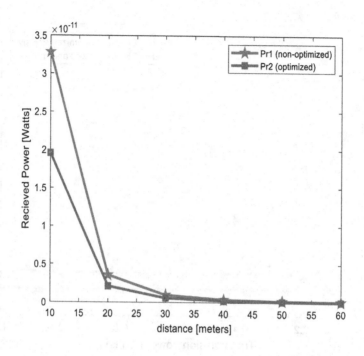

Figure 10. Comparative bar graph ofNon-optimized Received power ($P_r 1$) and optimized Power ($P_r 2$) varying with parameter distance (P_t)

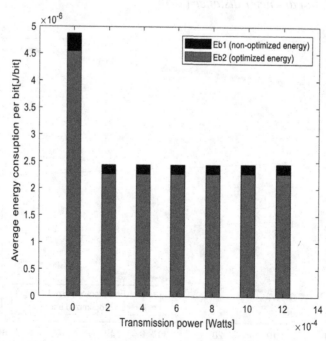

Figure 11. Comparative bar graph of Non-optimized Received power (P_r 1) and optimized Power (P_r 2) varying with parameter distance (d)

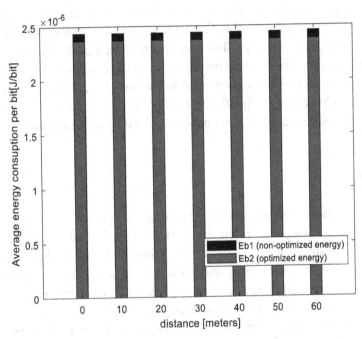

Figure 12. Comparative bar graph ofNon-optimized Received power (P_r 1) and optimized Power (P_r 2) varying with parameter distance (d)

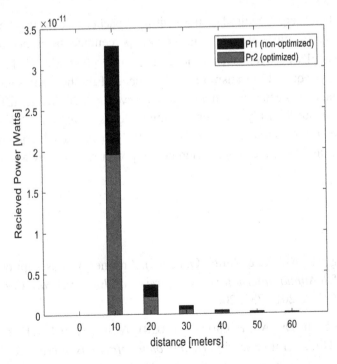

which clearly illustrates that the energy required for transmitting a packetsuccessfully from one node to other is decreased after optimization thus making the network efficient.

As the energy required for transmitting a packet successfully from one node to other is decreased the power received by the sensing node also decreases, the received power plotted against the distance between sensing nodes is given in Fig.10. The plot illustrates that as the distance is increased there is less difference between the received powers before and after optimization. As we have increased the efficiency of a WSN as compared to WSN given in (Abo-Zahhad et al., 2015) by optimization using PSO, the comparative bar graphs shown in Fig.10, Fig.11, Fig.12, clearly illustrates the reduction of energy required to transmit a packet successfully and power received by sensing node.

5 CONCLUSION

As the energy efficient communication systems are gaining importance day by day, the energy optimization in WSN has become a hot topic for research. Due to the rapid urbanization the energy management has become a challenge in smart cities, one of the key drivers for incorporating frameworks and making structures progressively smart is the energy management and saving that can be accomplished. Smart city is viewed as smart because of its innate insight in managing its assets and condition. It utilizes accessible data and communication advances particularly the IoT.

IoT inputs the necessary knowledge into essential structure of the city and helps make it smart, smart cities have monstrous potential in totally pivoting the operational effectiveness of a city and IoT is the specialized establishment behind these smart cities. The optimization using PSO for obtaining the optimal value of energy required to transmit a packet successfully between the sensing nodes has been studied. It has been found that this average energy gets reduced on optimization and thus increasing the overall efficiency of the wireless sensor network.

After optimization when energy required for transmitting a packet successfully between sensing nodes is plotted with respect to transmit power (P_t), the average percentage difference of energy required to transmit a packet successfully between sensing nodes is found to be 6.83%. It is also found that after optimization when energy required to transmit a packet successfully between sensing nodes is plotted with respect to the distance between sensor nodes (d) the average percentage difference of energy required to transmit a packet successfully between sensing nodes is found to be 2.75%. Due to the reduction in the energy required for transmitting a packet successfully there is a reduction in the power received by the sensing nodes, the average percentage difference of power received by a sensor node is found to be 40.42%.

REFERENCES

Abo-Zahhad, M., Farrag, M., Ali, A., & Amin, O. (2015). An Energy consumption model for wireless sensor networks. *IEEE 5th Annual International Conference on Energy Aware Computing Systems and Applications.* 10.1109/ICEAC.2015.7352200

Al-Garadi, Mohamed, Al-Ali, Du, Ali, & Guizani. (2020). A Survey of Machine and Deep Learning Methods for Internet of Things (IoT) Security. *IEEE Communications Surveys & Tutorials.*

Alghamdi, T.A. (2020). Energy efficient protocol in wireless sensor network: optimized cluster head selection model. *Telecommun Syst, 74*.

Amruta, M. K., & Satish, M. T. (2013). Solar powered water quality monitoring system using wireless sensor network. *IEEE Conference on automation, computing, communication, control, and compressed sensing*. 10.1109/iMac4s.2013.6526423

Chaturvedi, A., & Shrivastava, L. (2020). *IoT Based Wireless Sensor Network for Air Pollution Monitoring*. IEEE 9th International Conference on Communication Systems and Network Technologies (CSNT), Gwalior, India.

Gupta, S. K., Singh, R. K., & Sharan, S. N. (2016). An approach to implement PSO to optimize outrage probability of coded cooperative communication with multiple relays. *Alexandria Engineering Journal*.

Koohi, I., & Groza, V. Z. (2014). Optimizing particle swarm optimization algorithm. *IEEE 27th Canadian Conference on Electrical and Computer Engineering*. 10.1109/CCECE.2014.6901057

Li, J. (2019). A clustering based routing algorithm in IoT aware wireless mesh networks. *Sustainable Cities and Society*.

Liang, H., Yang, S., & Li, L. (2019). *Research on routing optimization of WSNs based on improved LEACH protocol. J Wireless Com Network*. doi:10.118613638-019-1509-y

Liu, W., Wang, Z., Liu, X., Zeng, N., & Bell, D. (2019). A Novel Particle Swarm Optimization Approach for Patient Clustering From Emergency Departments. *IEEE Transactions on Evolutionary Computation, 23*(4), 632–644. doi:10.1109/TEVC.2018.2878536

Papageorgiou, P. (2003). *Literature Survey on wireless sensor networks. Report*. University of Maryland.

Pinto, A. R., Poehls, L. B., Montez, C., & Vargas, F. (2012). *Power optimization for wireless sensor networks*. In *Wireless Sensor Networks - Technology and Applications*. IntechOpen. doi:10.5772/50603

Saha, D., Yousuf, M.R., & Matin, M.A. (2011). Energy efficient scheduling algorithm for S-MAC protocol in wireless sensor network. *International Journal of Wireless and Mobile Networks*.

Shafiabadi, M. H., Ghafi, A. K., & Manshady, D. D. (2019). *New Method to Improve Energy Savings in Wireless Sensor Networks by Using SOM Neural Network. J Serv Sci Res*. doi:10.100712927-019-0001-x

Chapter 5
Security and Privacy Issues in Smart Cities

Aditya Sam Koshy
Jamia Hamdard, India

Nida Fatima
Jamia Hamdard, India

Bhavya Alankar
Jamia Hamdard, India

Harleen Kaur
Jamia Hamdard, India

Ritu Chauhan
Amity University, India

ABSTRACT

The world is going through growth in smart cities, and this is possible because of a revolution of information technology contributing towards social and economic changes and hence endowing challenges of security and privacy. At present, everything is connected through internet of things in homes, transport, public progress, social systems, etc. Nevertheless, they are imparting incomparable development in standard of living. Unified structure commits to welfare, well-being, and protection of people. This chapter surveys two consequential threats, that is, privacy and security. This chapter puts forward review of some paperwork done before consequently finding the contributions made by author and what subsequent work can be carried out in the future. The major emphasis is on privacy security of smart cities and how to overcome the challenges in achievement of protected smart city structure.

DOI: 10.4018/978-1-7998-7541-3.ch005

1 INTRODUCTION

Kevin Ashton first used the term "Internet of Things" in 1999 (Ashton, n.d.). It was during his work at Procter & Gamble. Though the term IoT was coined just 21 years ago, the concept itself has been relevant since the 1970s. In (Domingue et al., 2009), IoT is defined as, "A world where physical objects are seamlessly integrated into a sophisticated network" and "Services are available to mingle with the help of these 'physical objects' over the Internet".

The IoT shows a lot of promise as a technology however, the concern lies in the security practices and height of privacy that this technology has achieved and can achieve. This paper focuses on different technicalities that stand out as drawbacks of IoTs and different techniques that have been employed, proposed or are being practiced currently. This paper is divided into 5 sections. Different aspects of security have been further discussed with different approaches that people all over the world have taken to counter this short-coming. The safety features of IoTs are very well covered under three aspects as shown in figure below (Ngu et al., 2016):

Figure 1. Three Aspects of Safety

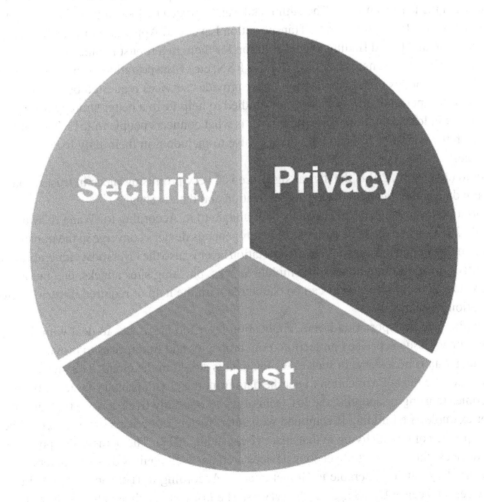

The security is often defined in terms of 3 traits- Confidentiality, Integrity and Availability as in (Jang-Jaccard & Nepal, 2014). Confidentiality ensures that the data is safe and no one gets it except the one that the data is meant for. Integrity ensures that data is handled or dealt in an honest manner by the desired people. And availability means the system is strong and stable enough to cater all the needs and function properly for all. In (Lopez et al., 2009), more properties of security are discussed including forward and backward secrecy, etc.

The privacy characteristic ensures anonymity and seclusion from the entities that one doesn't wish to share the data with. This characteristic provides the client or user the appropriate safety that their information is not viewed by undesirable people like hackers.

'The trust is the characteristic that deals with developing a sense of accountability in the service providers with respect to the data or service pertaining to any client'. The trust factor plays vital role in getting people on board with the said technology with the recognition that they are protected by the enterprises they seek service from. The trust factor can be defined in terms of traits like authentication and authorization in addition to accountability. According to (Burrows et al., 1989), the authentication ensures that the data that is sent through the server is sent by a legit person, whereas (Cirani et al., 2015), states that authorization ensures that the data is received by an authorized person for processing.

In paper (Frustaci et al., 2018), the authors have discussed the present and future challenges to the security concern of the IoT technology. The authors take an approach by focusing on the different layers in the model of IoT viz., Perception Layer, Transportation Layer and Application Layer. This approach helps in focusing on threats and finding a valid solution for them in a robust manner.

Perception layer includes sensors like GPS, RFIDs, RSN, etc. Transportation layer is the medium for the devices in perception layer and Application Layer provides services requested by the devices. The certain vulnerabilities pertaining to each layer are studied to help form a better strategy for countering the element of risk in IoT adaptation. Perception layer is what connects people to IoTs. To make people comfortable with the notion of IoT and help them evolve to include it in their daily lives, Trust Factor plays a vital role.

According to (Jing et al., 2014) the digital signature is a very promising piece in trust management system to claim data authentication, device authentication and data exchange.

In transportation layer we arrive to issues that are unique to it. According to (Wang & Wang, 2011) & (Zhang et al., 2013), when the wide varieties of heterogeneous devices converge in the transport layer, security problems are bound to happen. The transportation layer is also the layer most vulnerable to DDos attacks (Yi, 2010), in addition to attacks like middle attacks, phishing sites attacks, etc. but according to (Zhang & Wang, 2006), these attacks can be countered with the help of required intrusion detection and authentication mechanisms.

Like perception and transportation layers, application layer too has issues to deal with. In application layer the primary issue is privacy protection. For companies and Enterprises to provide services to clients, obtaining data is necessary; in such cases the data is stored in the cloud. This makes the cloud and the ones' that have the authentication to get the data from the cloud, primary target of the hackers.

When it comes to using or applying the IoT technology to our daily lives, a new spectrum of issues opens up. For example, if a building is equipped with smart meters or perhaps 'cams', it could be very easy to invade privacy of people living in that space (Ning et al., 2017). The botnets also pose a serious threat to IoT devices. Since the heterogeneous devices aren't as well equipped as computers or mobile devices, they are very much vulnerable to Botnet attacks. According to (Hutson, 2017), Mirai botnet can infect devices, and cause DDoS against the servers. The instances such as sabotage of traffic lights can be imagined to be very disruptive towards the society and dangerous too. The collection of sensitive data is another concern. Since the devices in a household could be used for monitoring, it is easy to understand that there are chances that sensitive information are being collected and could be used to pinpoint the location or whereabouts of any potential target (Petrolo et al., 2017). The other forms of risk could be the breach of healthcare information and lifestyle and habits.

More challenges are faced in making IoT usable by the public. Few of them are mentioned:

Connectivity: The whole idea behind IoT is to be able to bring the devices and people all over places together in such a manner that devices are able to interact with each other for better facility, convenience and help. This would on be possible if the devices have a good server acting as a medium for the exchange. Making sure that the devices even in most remote places are able to send and receive information is a priority.

Heterogeneity: The IoT consist of many devices communicating with each other over network. All these devices are of different types which means, a single scheme of security model would find it hard to accommodate all the devices under its blanket of safety. With the differences in the types, functions and vulnerabilities of every device, it is needless to say that we need to devise proper techniques to ensure their safety.

Large Networks: A large scale network consist a large no. of devices which means there are more devices that are to be accounted for. This can be a tough task because sheer no. of devices can be overwhelming.

Resource Constraints: Given the fact that the IoT is serving a purpose that is to make information readily available and have a more systematic approach to things using certain devices, it is only limited. For example, devices like CCTV cameras can't be replaced by a professional photoshoot standard camera with high end specs, that won't exactly be the resourcefulness that we intend to achieve.

Maintenance of Devices: The devices in an IoT network are but electronic devices. They do need proper attention to make sure of their proper functioning. This can be a very difficult feat to achieve since in IoT devices are of wide variety. The need to make sure of devices sufficient working also demands it to be able to withstand the environmental conditions that it is supposed to reside in. For instance, the Traffic Cams need to be durable enough to function properly on Sunny days as well as rainy days without succumbing to the environmental conditions.

1.1 Our Contribution

The following contributions are made in this paper:

1. Study about Smart City Ecosystem, Opportunities, Issues and Challenges
2. Study of various state of art literature covering Security and Privacy of Smart Cities
3. Analysis of State of the Art about Smart City technology
4. Suggested some of the best practices to ensure privacy and security in Smart Cities

This paper covers the previous works that have been done in the field of Smart Cities. In section 2 we have discussed a few approaches made in different literatures in the field of Smart Cities. In section 3, a tabular comparison discussing different aspects and techniques suggested in the literature have been discussed. In Section 4, we have pointed out the best practices that can help in making Smart cities more safe and secure. In Section 5, provides the conclusion of the paper.

2 RELATED WORKS

Security Challenges are so certain about smart cities wherefore researchers should find solutions for them. In this section, discussion is done on foregoing research in the field security and privacy of smart cities along with what are the challenge smart cities are witnessing. In (Batty et al., 2012), author talked about factors how to make cities smarter and display what are the risks smart cities are facing. In meantime all over 2014, writers begin to bothered about issues related to smart cities such as security and privacy along with challenges of smart city' corresponding to it (Elmaghraby & Losavio, 2014), put up model mainly focus at major smart cities services and challenges faced by them. In 2017, (Usman et al., 2017) incorporate uniform substitution of permutation method and Fiestel method for building novel process so as to eliminate issues face by smart cities and named it secure IOT (SIT). The (Petrolo et al., 2017) talk about integrating cloud computing alongside Internet of Things for starting a modern application termed as Cloud of Things (CoT), after it challenges are reviewed along with benefits of smart city, in meantime author explain how to use Cloud of Things (COT) technology with services of smart cities. Conveniently by 2018, (Ainane et al., 2018) proposed data transmission between components of network as well as put forward IOT should indulge in effectual regulations of threats, acknowledging them and evaluation of warnings in smart city structure. Analysts have come across scientific determination as for how challenges can be overcome facing by smart cities alongside how privacy can be achieved. The (Abosaq, 2019) examines affairs called out in agile conurbation like validation, clandestinely, faith, specific remoteness, and safety of data, assess control, secure middleware, and policy implementation. Writer proposed simulated smart city model should be well connected with required Communication channels from which data is produced for sum total of sensors. Study of (Awad et al., 2019) contend about compromising privacy easily as there is high degree of interconnection between people, devices and sensors, thus accentuate that there is need to fully secure this data. In(Meng et al., 2018) researchers affirms about challenges IoT devices are facing like spoofing attacks, jamming, unauthorized access, etc., through which probity of user data can be negotiated. He further said there are numerous solutions through which security of IoT devices can be bring off just by implementing various security checks on IoT devices. The author (Siby et al., 2017) discussed about privacy threats that are arising in today's world because of infiltrating IoT technologies and connected networks. He suggested businesses and organizations should have a proper check over the scanning and monitoring of IoT devices so any vulnerability or attack can be figure out privacy is maintained, data breaches does not circulated. He contributed with the help of Traffic inceptors and analyzers cyber-attacks can be identified and resolved by looking into it. Author emphasis in (Hassan, 2019) about research progress carried out in the field of IoT security. He remarked various simulation tools and modelers hold up the research. However, if IoT devices breakdown the resulting results will be acute. Researchers in (Leloglu, 2016) suppose in spite of boundless assistance, IoT devices are facing numerous challenges. Two threats that are at glance are Cyber security and Privacy issue. There two are of worry for both public and private organizations. IoT technologies vulnerabilities have shown eminent cyber security threats as interrelation of network and Internet of Things are accessible from untrusted network inconsequence, security solution is required and mandate. Further it was remarked standards, protocols and principles should be followed while implementing IoT Cyber Security Framework. In (Ali et al., 2018), author reports about ongoing IoT Cyber Security solutions. Basic security methods have been evaluated. It was brought out that organizations are doubtful whether they are in state to develop right solutions for cyber security or not.

3 STATE OF THE ART IN SMART CITY SECURITY

Following is an elaborative table that discusses State of the Art in the field of Smart City Security. Their limitations have also been drawn out to better facilitate the comparison amongst them.

Table 1. State-of-the-art in field of Smart City Security

S.No	Ref	Contribution	Limitations	Remarks
1	(Cui et al., 2018)	This paper mentions different layers of IoT Architecture and wide range of its applications that include Smart Transportation, Utilities, etc. the authors also point out the characteristics and Security requirements of an IoT network.	The issues with the mentioned in this work include: Botnet Attacks, Driverless Cars, Privacy Issue, and AI Threats.	The Rapid development of various devices that comprise IoTs, have opened up several paths to further our progress in the given field. This also calls for enhanced techniques and measures for privacy and security of individuals that are part of the Smart Cities.
2	(Ismagilova et al., 2020)	The study proposes authentication architecture for IOT devices. This architecture is based on RFID application which suits for smart environment conveniently.	Vulnerability for smart city' developer and planners are the techniques such as Blockchain build structure alongside use of Artificial intelligence system as they are intrinsic for framing of human's in nesting of systems so how to provide security to the people and privacy too.	Author proposes upcoming work should regard how lawful structure can be integrated for resolving faith deficit of citizens in structure of smart city. This research put up that finding to be carried out in solving existing vulnerabilities which will yield to resourcefulness of smart cities.
3	(Kabir, 2020)	This Paper rounds out best IoT security Techniques that are available currently like - IoT Authentication, IoT Encryption, IoT Security Analytics, etc. The author also mentions Wireless Security techniques that are applicable: Wired Equivalent Privacy, WiFi Protected Access, etc.	The protection in IoT focuses on 2 aspects. One is protection of data and the other is Security and integrity of devices present in the network. Because it is such an inter-connected framework of devices, it requires as much attention to make sure of the safety and security	The IoTs have introduced a new path for development of service sector. Embracing the newer technologies is always appreciated but a thorough study and research would only help us to move in a more sustainable and safer way. IoT
4	(van Zoonen, 2016)	Author labels the information collected by the devices on two factors: i)Data ii)Purpose The Data can be Personal or Impersonal, while the purpose can be Surveillance or Service. Based on this Schematic, 4 types of data can be labeled. The author explains the use of privacy framework with the use of 3 examples: smart waste technologies, predictive policing, social media monitoring	Limitations and concern for privacy of data rise from different factors that play major role in recognition of the limitations and ways to overcome them: -Kind of Data -Purpose of Data -Who collects Data	To get a better grip on concerns regarding data privacy, the author tends to draw attention on types of data and how they are collected. The perception of people in the smart city is also vital key for better understanding.
5	(Hossain et al., 2015)	This paper focuses on security problems of IoT environment. Different components are considered in this paper to study the challenges that occur in building an IoT network. The author also discusses Different attacks that are main threats: Access Level Attacks, Location based, etc.	IoT Network is highly resource constrained. Different factors that contribute to drawbacks of an IoT system are based on software, network, etc.	IoT being such a technology that brings a number of heterogeneous devices together, opens up the world to newer possibilities. This being the case, it is also to be noted that the vulnerabilities are not to be undermined. There are several cases discussed in this paper that give us an idea about the streams of IoT that need attention for further development.
6	(Ali, 2019)	The author draws our attention towards various IoT security techniques such as Blockchain-based IoT Security, IoT Security Based on AI Techniques, Security of Cloud–based IoT Environment, etc.	The Challenges faced by IoT security techniques are many. This paper mentions all of the challenges as follows: Privacy, Device Resource Constraint, Hardware/Firmware vulnerabilities, Heterogeneity of devices,etc.	IoT is a very fast growing field that has very promising aspects. The challenges in this field of technology are vast too. This paper works as a nudge in the direction of opportunities that lay in resolving issues in this field.

continues on following page

Table 1. Continued

S.No	Ref	Contribution	Limitations	Remarks
7	(Tiwary et al., n.d.)	This paper focuses on various aspect of an IoT environment, including architecture, components, etc. The author proposes that hardware components such as sensors, actuators, Middleware components like database or cloud for storage and data analytical tools and Visualization via varied applications comprise the vital components of an IoT	The IoT being as promising as it is does come with as many factors limiting its application: Security, Privacy, Complexity, Flexibility and Compliance	The features of IoT do provide a path for further enhancement in a global manner. The people are already able to interact and communicate with others far away, IoT furthers this campaign and draws attention towards a possibility of objects being able to interact and steering lifestyle into a new direction.
8	(Se-Ra, 2017)	The authors have tried to determine the security issues in the IoT based on their 3 characteristics: Heterogeneity, Resource Constraint & Dynamic Environment. The authors have focused on working out Security issues by studying the Key Elements of the IoT structure including the user and cloud.	This paper discusses the challenges in Security of IoT based on each Element of it are mentioned as follows: Privacy, Authentication, Encryption of data, Proper Security Regulations, ID-Password integrity Physical threats to devices, Trust Access Control, Middleware & Storage	This paper has tried to shed light on the security issues by considering the 3 basic characteristics of the IoT: Heterogeneity, Resource Constraint, and Dynamic Environment. IT paves way for further analysis of limitations and challenges in this field.
9	(Butt & Afzaal, 2019)	Author has proposed that security of smart cities is characterized on two factors namely operational security and data security.	In present scenario systems are being deployed without testing for cyber security and attackers can smartly break the encryption with cold boot attack' and using other channels.	Authors recommend the use of compressed IPsec achieve encryption between smart cities and IPv6 to achieve a level of privacy necessary for the functioning of the entire system.
10	(Farahat & Tolba, 2019)	The author has present authentication encoded encryption technique which may settle few security challenges IoT are facing in smart cities.	The considerable issue with smart cities is how to assure sensors data from being purloin from it source to destination.	Author has discussed only authorized people can ingress the data by authentication process and in the course of time focal point is to integrate projected system in smart city services.
11	(Al-Turjman et al., 2019)	The author has discussed Information Centric paradigm to be contemplate as a replacement for IP based network in Smart Cities'.	Application that are cloud based are insufficient which can be overcome by fog based computing but at the same time they are not appropriate to resolve constraint in IOT devices	The authors scrutinize problems that arise in a smart city application, mostly pertaining to the safety aspect.
12	(Barun, 2018)	Author proposes, when determining way for protecting citizen privacy in smart cities data set, the conception of differential privacy should be considered.	The challenges are how to preserve privacy of high dimensional data, how to network that are more prone to attacks and attenuate failures cascading through the smart networks.	Author discusses in order to convince residents computational trust should be establish with in the network, doing so will probably provide assurance for Communication and all parties will be obliged to abide by the regulations of smart city.
13	(Kamble & Bhutad, 2018)	Security Techniques that are focused upon in this paper are classified into different layers: Perception Layer- Authentication is achieved through Cryptographic Hash Algorithms, etc. Network Layer-Proper Authentication, Appropriate routing algorithms help build up privacy, etc. Middleware and Application Layer-Security is ensured with help of different techniques like Public Key Encryption, etc. & For detecting intrusions into the network techniques like anomaly detection, etc. are used.	This paper discusses Challenges faced by IoT based on each of its aforementioned layers Perception Layer Challenges Unauthorized Access to Tags, Node Capture Attacks, etc. Network Layer Challenges Spoofing Attack, Sinkhole Attack, Denial of Service (DoS) Attack, etc. Application Layer Challenges Phishing Attack, Malicious Virus/worm, etc.	With newer enhancements in the field of IoT, better safety techniques are also needed to be developed. The author proposes that Data Analytics, Artificial Intelligence and Machine Learning will play vital role in future.

continues on following page

Table 1. Continued

S.No	Ref	Contribution	Limitations	Remarks
14	[41]	This paper perceives IoT through 3 different orientations: Internet oriented, Things oriented, & Semantic oriented. The authors propose security measures that ensure low risk usage of this technology: Encryption, Confidentiality, Authentication, Authorization, Certification and Access Control	The limitations that IoT faces are discussed in this paper with respect to their layers: **Perception Layer** Low computational power, limited capacity of perception devices, etc. **Transportation Layer:** Sybil Attack, Sinkhole Attack, etc. **Application Layer:** Malicious Code Injection, Denial-of-Service (DoS) Attack, etc.	The IoT is a very promising Technology that would create a world where everything is connected. This would require a huge backing from security point of view because of vast data that is would handle including all the devices in the network. The security issue studied at each layer is a very effective approach for better understanding the challenges and providing a solution.
15	(Elmaghraby & Losavio, 2014)	The author mentions work done by IBM, CISCO and SIEMENS in this field. Their findings have been mentioned below: IBM: Towards a smarter planet -Intelligent. -Interconnected. - Instrumented. SIEMENS: Sustainable cities are served by -Intelligent traffic solutions. -Green Buildings. -Water Management. CISCO: Smart Interconnected Communities -Improved Citizen Interaction. -Connected Residential Communities.	Security Issues: Illegal access to Information and Physical Disruption and several others. Privacy Issues: Digital citizens are more and more instrumented, etc. Technology Issues: -Use of electronic data for marketing and targeting purpose.	The Smart cities are emerging as newer and better lifestyle options. Adopting such technologies is appreciated but the challenges pertaining to them should also be studied thoroughly for a better tomorrow.

4 Best Practices for Achieving Privacy Preserved and Secured Smart City

In this section we discuss, different practices that help in making Smart Cities digitally safer. The diagram shows various solutions that would help in reducing vulnerabilities of the Smart City.

Blockchain for Transparency: System access can be gained by hackers by taking advantage of edges devices incorporating switches and routers. Blockchain is used to protect devices from attack and vulnerabilities. In this technology a network cannot be controlled by centralized authority so attacking a device is not easy.

Consider Patching and Remediation: IoT devices must be considered by businesses as per their capability for patching and remediation. Devices should be evaluated and other business code must be changed overtime for securing devices.

Proper Protocols: There is a need for protocols to be introduced, which would command or mediate the way interaction takes place over the IoT Network. The idea is to create a set of protocols that are analogous to internet protocols over the internet.

Authentication & Confidentiality practices: Perhaps to make IoTs safer, these two concepts play the most vital roles. The authentication would ensure the data is handles by a legit person and confidentiality would help in keeping the information shared anonymous.

Prediction Algorithms and Intrusion Detection: The IoT is very much vulnerable to attacks as discussed earlier. Employing proper prediction algorithms and intrusion detection mechanisms is an important job so as to prepare for any future attacks from hackers.

Figure 2. Diagram showing different practices that are recommended for safer Smart City experience

Data Transparency: Whenever a user enters some data or chooses to share some information through IoT, they should have the right or flexibility of sorts to decide who all get to see the data or who doesn't. The user can decide it better if the data they are entering is sensitive or not, so they are the better judge when it comes to deciding who should see the data and who not.

IoT Traceability: In case of any cybercrime, the authority should be able to traceback the digital fingerprints so as to reach the point of breach in order to catch the criminals behind it. It is to be noted that this is not a recommendation towards installing a backdoor into certain systems, but only a way to record the steps or ins and outs of a safe digital gateway in many ways similar to call logs. This is to be programmed carefully so as it does not end up breaching the privacy of individuals.

Educating Crowds: The people in coming age, should know and understand the impact of digitization and absorb it in a positive manner while getting along. They need to learn ways to make sure they don't end up sharing anything through IoTs that may compromise their safety or well-being. The awareness should be spread regarding maintaining anonymity and keep their data or information confidential.

Proper disposal of Devices: When device becomes obsolete, users tend to just throw them away without discarding private data. Later on it can reveal personal information of user within IoT ecosystem. So "discard, recycle or destroy (DRD)" policy should be followed as it can regulate which device should be disposed of and how.

5 CONCLUSION

This paper lays emphasis on conceptual structure of smart cities published work and the research carried out before, thereupon focusing on issues like security of cities with privacy too, besides what are the effects they can build up on the process. Much of the challenges has been discussed and studied about so a valuable conclusion can be drawn and what can be the elements contributing for the achievement of factors like insurance (security), privacy in smart city'. In literature review, a number of premises have been developed and discussed. The study of existing writings draw attention to that cities are lacking in quantitative and qualitative analysis of data in the espousal of smart cities. Well-known threats existing in architecture of smart cities are mainly information theft, unauthorized access of data, replication of data, injecting virus-based attack to the devices, human factors etc. Moreover, it reviews issues related to security of smart city, privacy of smart city and how to prevail over challenges. In this paper issues have been discussed and what are the available solutions to get better of the challenges for the achievement of secured smart city structure.

REFERENCES

Abosaq, N. H. (2019). Impact of privacy issues on smart city services in a model smart city. *International Journal of Advanced Computer Science and Applications*, *10*(2), 177–185. doi:10.14569/IJACSA.2019.0100224

Ainane, N., Ouzzif, M., & Bouragba, K. (2018). Data security of smart cities. *ACM International Conference Proceeding Series*. 10.1145/ 3286606.3286866

Al-Turjman, Zahmatkesh, & Shehroze. (2019). *An overview of security and privacy in smart cities' IoT communications*. Academic Press.

Alandjani, G. (2018). Features and potential security challenges for IoT enabled devices in smart city environment. *International Journal of Advanced Computer Science and Applications*, *9*(8), 231–238. doi:10.14569/IJACSA.2018.090830

Ali, M. A. (2019, December). IoT Security Evolution: Challenges and Countermeasures Review. *International Journal of Communication Networks and Information Security*, *11*(3).

Ali, S., Bosche, A., & Ford, F. (2018). *Cybersecurity Is the Key to Unlocking Demand in the Internet of Things*. Bain and Company.

Ashton. (n.d.). That 'Internet of Thing' thing. *RFID Journal*.

Awad, A. I., Furnell, S., Hassan, A. M., & Tryfonas, T. (2019). Special issue on security of IoT – enabled infrastructures in smart cities. *Ad Hoc Networks*, *92*, 101850. Advance online publication. doi:10.1016/j.adhoc.2019.02.007

Barun, B. C. M. (2018). Security and privacy challenges in smart citie. *Sustainable Cities and Society*, *39*, 499–507. doi:10.1016/j.scs.2018.02.039

Batty, M., Axhausen, K. W., Giannotti, F., Pozdnoukhov, A., Bazzani, A., Wachowicz, M., Ouzounis, G., & Portugali, Y. (2012). Smart cities of the future. *The European Physical Journal. Special Topics*, *214*(1), 481–518. doi:10.1140/epjst/e2012-01703-3

Burrows, M., Abadi, M., & Needham, R. M. (1989). A logic of authentication. *Proceedings of the Royal Society of London A: Mathematical, Physical and Engineering Sciences*, *426*(1871), 233–271.

Butt, T., & Afzaal, M. (2019). *Security and Privacy in Smart Cities: Issues and Current Solutions.* doi:10.1007/978-3-030-01659-3_37

Cirani, S., Picone, M., Gonizzi, P., Veltri, L., & Ferrari, G. (2015). IOT-OAS: An oauth-based authorization service architecture for secure services in iot scenarios. *Sensors Journal, IEEE, 15*(2), 1224–1234. doi:10.1109/JSEN.2014.2361406

Cui, Xie, Qu, Gao, & Yang. (2018). Security and Privacy in Smart Cities: Challenges and Opportunities. *IEEE Access.*

Elmaghraby, A. S., & Losavio, M. M. (2014). Cyber security challenges in smart cities: Safety, security and privacy. *Journal of Advanced Research, 5*(4), 491–497. doi:10.1016/j.jare.2014.02.006 PMID:25685517

Farahat, I. S., & Tolba, A. S. (2019). *Security in Smart Cities: Models, Applications, and Challenges.* Springer.

Frustaci, Pace, Aloi, & Fortino. (2018). Evaluating Critical Security Issues of the IoT World: Present and Future Challenges. *IEEE Internet of Things Journal, 5*(4).

Haller, S., Karnouskos, S., & Schroth, C. (2009). The Internet of Things in an Enterprise Context. In J. Domingue, D. Fensel, & P. Traverso (Eds.), Lecture Notes in Computer Science: Vol. 5468. *Future Internet – FIS 2008. FIS 2008.* Springer. doi:10.1007/978-3-642-00985-3_2

Hassan, W. H. (2019). Current research on Internet of Things (IoT) Security: A Survey. *Computer Networks, 148,* 283–294. doi:10.1016/j.comnet.2018.11.025

Hossain, Fotouhi, & Hasan. (2015). Towards an Analysis of Security Issues, Challenges, and Open Problems in the Internet of Things. *IEEE World Congress on Services.*

Hutson, M. (2017). A matter of trust. *Science, 358*(6369), 1375–1377. doi:10.1126cience.358.6369.1375 PMID:29242328

Internet of Things. (n.d.). In *Wikipedia.* www.wikipedia.com

Ismagilova, E., Hughes, L., Rana, N. P., & Dwivedi, Y. K. (2020). Security, Privacy and Risks within Smart Cities: Literature Review and Development of a Smart City Interaction Framework. *Information Systems Frontiers,* 1–22. doi:10.100710796-020-10044-1 PMID:32837262

Jang-Jaccard, J., & Nepal, S. (2014). A Survey Of Emerging Threats In Cybersecurity. *Journal of Computer and System Sciences, 80*(5), 973–993. doi:10.1016/j.jcss.2014.02.005

Jing, Q., Vasilakos, A. V., Wan, J., Lu, J., & Qiu, D. (2014, November). Security of the Internet of Things: Perspectives and challenges. *Wireless Networks, 20*(8), 2481–2501. doi:10.100711276-014-0761-7

Kabir. (2020). *An overview of the Internet of Things (IoT) and IoT Security.* Research Gate.

Kamble, A., & Bhutad, S. (2018). Survey on Internet of Things (IoT) security issues & solutions. *2nd International Conference on Inventive Systems and Control (ICISC),* 307-312. 10.1109/ICISC.2018.8399084

Leloglu, E. (2016). A review of security concern in Internet of Things. *J. Comput. Commun., 5*(01), 121–136. doi:10.4236/jcc.2017.51010

Lopez, J., Roman, R., & Alcaraz, C. (2009). Analysis of Security Threats, Requirements, Technologies And Standards In Wireless Sensor Networks. In Foundations of Security Analysis and Design V. Springer. doi:10.1007/978-3-642-03829-7_10

Meng, Y., Zang, W., Zhu, H., & Shen, X. S. (2018). Securing consumer IoT in the smart home: Architecture, challenges and countermeasures. *IEEE Wireless Communications*, *25*(6), 53–59. doi:10.1109/MWC.2017.1800100

Ngu, Gutierrez, Metsis, Nepal, & Sheng. (2016). IoT Middleware: A Survey on Issues and Enabling Technologies. *IEEE Internet of Things Journal*.

Ning, Z., Xia, F., Ullah, N., Kong, X., & Hu, X. (2017). Vehicular social networks: Enabling smart mobility. *IEEE Communications Magazine*, *55*(5), 16–55. doi:10.1109/MCOM.2017.1600263

Petrolo, R., Loscri, V., & Mitton, N. (2017). Towards a smart city based on cloud of things, a survey on the smart city vision and paradigms. *Transactions on Emerging Telecommunications Technologies*, *28*(1), e2931. doi:10.1002/ett.2931

Se-Ra, O. (2017). Security Requirements Analysis for the IoT. Academic Press.

Siby, S., Maiti, R. R., & Tippenhauer, N. O. (2017). IoTscanner: Detecting privacy threats in IoT neighborhood. *Proceedings of the 3rd ACM International Workshop on IoT Privacy, Trust and Security*, 23-30.

Tiwary, Mahato, Chidar, Chandrol, Shrivastava, & Tripath. (n.d.). Internet of Things (IoT): Research, Architectures and Applications. *International Journal on Future Revolution in Computer Science & Communication Engineering*, *4*(3).

Usman, M., Ahmed, I., Aslam, M. I., Khan, S., & Shah, U. A. (2017). *SIT: a lightweight encryption algorithm for secure internet of things.* arXivPrepr. arXiv 1704.08688

van Zoonen, L. (2016). Privacy concerns in smart cities. *Government Information Quarterly*, *33*(3), 472–480. doi:10.1016/j.giq.2016.06.004

Vashi, S., Ram, J., Modi, J., Verma, S., & Prakash, C. (2017). Internet of Things (IoT): A vision, architectural elements, and security issues. *International Conference on I-SMAC (IoT in Social, Mobile, Analytics and Cloud) (I-SMAC)*, 492-496. 10.1109/I-SMAC.2017.8058399

Wang, Z. L., & Wang, F. H. (2011). *Introduction to the internet of things engineering.* Mechanical Industry Press.

Yi, K. M. (2010). *Preliminary study of IoT security.* Internet Police Detachment of Public Security Bureau in Taiwan City.

Zhang, G. G., Bi, Y., & Li, C. (2013). Massive internet data security processing model research. *Small Microcomputer System*, *34*(9), 2090–2094.

Zhang, L., & Wang, Z. (2006). Integration of RFID into wireless sensor networks: architectures, opportunities and challenging problems. In *Proceeding of the IEEE fifth international conference on grid and cooperative computing workshops GCCW '06* (pp. 463–469). 10.1109/GCCW.2006.58

Chapter 6
A Conceptual Model to Next-Generation Smart Education Ecosystem

Palanivel Kuppusamy

iD https://orcid.org/0000-0003-1313-9522
Pondicherry University, India

Suresh Joseph K.
Pondicherry University, India

ABSTRACT

Transforming IT systems educational applications has become imperative in a rapidly evolving global scenario. Today, educational organizations have to provide transparent, confident, secured information and quality data for monitoring and advanced predictive capabilities services to society. Educational organizations have to meet these objectives consistently during typical and crisis scenarios. Modern educational applications are integrated with social network sites, sensors, intelligent devices, and cloud platforms. Hence, data management solutions serve as the basis for educational organizations' information needs. However, modern technologies demand a re-engineer of these platforms to meet the ever-growing demand for better performance, scalability, and educational organizations' availability. This chapter discusses the challenges inherent to the existing educational data system, the architectural methods available to address the above challenges, and the roadmap for building next-generation educational data ecosystems.

I. INTRODUCTION

Digital transformation is a combination of modern tools and developments leveraged to solve business difficulties. It allows rethinking about how business institutions use technology, stakeholder, and techniques to change business improvement. Emerging technologies and revamped processes are crucial to any digital transformation.

DOI: 10.4018/978-1-7998-7541-3.ch006

Today, the education sector is continuously needed to be assessed and altered to follow business-changing trends (Christos, 2016). Higher Educational Institutions (HEIs) need an integrated view across the institutions to provide techniques to govern and manage the educational data (Sonali, 2019). This integrated view combines technical, academic operations, and processes across the data supply-chain to analytics applications.

The emerging technologies currently used in educational sectors are mobile technology, video conference, remote access systems, educational platforms, and services. These emerging technologies have drastically transformed the way education is conducted. Artificial Intelligence (AI), blockchain, Big data, cloud computing, edge computing, Internet of Things (IoT), Mobile Computing (3G/4G/5G), Robotic Process Automation, etc., are examples of convergence of technologies used in educational sectors. These technologies lead to the design of a smart educational environment.

Smart learning environments drive data-driven systems in educational sectors. Big data technology brings together data management challenges – the volume of data, various data, and complexity. The Internet of Things (IoT) interconnects sensing and actuating devices and their data with the Internet. Few concepts from the IoT perspective are smart cities, smart farming, smart agriculture and precision agriculture, smart industry 4.0, smart education, intelligent analytics, etc.

The IoT supports modern technologies, sustainable infrastructure development, enhances quality, and improves smart education systems' efficiency. The management supports educational systems in a highly integrated with the delivery system in an elegant manner. The IoT manages the real-time interactions between people, products, and devices in the educational environment. These emerging and converging technologies work in the interest of collaborating as part of an educational ecosystem.

The educational ecosystem's stakeholders or end-users are students, teachers, researchers, recruiters, parents, evaluation specialists, content creators, recruiters, governing bodies, other educational institutions, and decision-makers. They interact with them and take the necessary initiatives to improve teaching and learning in educational institutions. These technologies generate vast amounts of educational data ranging from individual access to institutional activities.

The massive educational data includes students' usage data, students' interaction data, learning management systems (LMS) data, students' learning activity data, and courses information data. For example, the course data consists of a curriculum informations such as learning objectives, syllabuses, learning activities, assessment results, social network activities, students' browsing history, and evaluation courses. The other educational service data are administration, governance, policies, standards, academic excellence, processes enhancement, and actions.

The educational systems can use these data and exploit them to improve performance, quality, better decision making, and recommendations. Data scientists and analysts, educational administrators, and governing bodies collect these massive data insights and lean increasingly using AI, ML, and DL technologies.

MOTIVATION

Today, educational institutions are extensively data-driven, and they need to preserve their data for several years for analysis with business intelligence (BI) tools and modern technologies. Educational institutions need to integrate their tasks with data management strategies, where the data from diverse systems into a central repository.

The educational administrators can conduct testing on these educational data for simulating multiple scenarios. These simulations typically generate a vast volume of (e.g., terabytes or petabytes) of data. Analysis of enormous amounts of data (i.e., structured and unstructured data) is key to assessment and monitoring. Educational institutions can use real-time analytics on the data from video surveillance, Weblog, and social media sites. This real-time analysis help identifies thefts and improves security in learning environments.

Educational institutions must therefore address the data challenges about volume, velocity, and variety. The issues and problems related to data ecosystems in the modern education system are coordination, collaboration, data management, a framework, emerging technology, and data security.

PROBLEM STATEMENT

Traditional data ecosystems comprising a presentation, data store, data warehouse, and data mart have coexisted with Big Data technologies. It requires massively scalable but inexpensive data storage solutions. However, most educational organizations are now building and developing *advanced data platforms* that utilize emerging analytics technologies. Educational organizations have started to create landing and processing zones for enterprise-wide data, external data feeds, and unstructured datasets.

The Data Warehouse is not a perfect solution for integrating unstructured data from sensor devices, social media networks, weblogs, etc., installed in the educational learning environment. The data warehouses with relational technologies for traditional structured data poses challenges to real-time decision making.

The digital transformation in education produces a learning ecosystem with learning data on every aspect of the educational institutions and their information systems. Learning data ecosystem presents the challenges in smart system design, which require dealing with data management in large-scale environments supporting openness, flexibility, and scalability with a minimum cost and time.

The smart education system's key challenge is to extract useful and precise insights from the massive data generated by various educational applications and make valuable decisions and recommendations to the stakeholders' educational sector.

THE SOLUTION

In educational data ecosystems, vast and diverse data move among various applications and actors in a supply-chain manner. These data are heterogeneous in nature. They can form different behaviors around educational organizations and technology platforms across various other organizations.

Hence, it is projected to provide a data management solution to a smart education system that needs to support knowledge flows in educational information systems. Integrating modern data analytical approaches with education sectors can investigate and improve educational processes and curriculum and improve educational content quality. These approaches address a majority of the data management challenges.

1. They provide a solution for hybrid data types where data can be processed either in real-time or in batch mode.

2. They provide a better solution to change the current Extract, Transform, and Load (ETL) format to the Extract, Load, and Transform (ELT) format. They can also be used to perform data transformation and aggregation processes. The aggregated data then moves to the Data Warehouse appliances within the traditional Data Warehouse.

3. These approaches offer an empirical analytics platform that effectively analyzes structured, unstructured, and semistructured data and seamlessly integrates with the existing analytics tools.

Data analytical approaches ensure seamless integration with the existing technology landscape and efficient use of expensive Data Warehouse appliances. They provide a comprehensive roadmap to assimilate the traditional data system into the next-generation data ecosystem. They support existing functionalities and create new ones while keeping the cost of services low.

Objective

This chapter discovers a learning data ecosystem in building a smart education system to support intelligent learning environments. The chapter examines different education data ecosystem elements that are serious to data sharing and management challenges. The conceptual model enables the education system flexibly to host, accomplish, and enhance many services to next-generation smart education environments.

The contributions are as follows:

- It introduces a next-generation learning data ecosystem that supports computing, data analytics, and resource transfer across devices, access, and tiers.
- It builds a conceptual model that includes data flows, i.e., creating and delivering the datasets to the stakeholders.
- It defines the data analytics architecture that supports users' demands with less operational costs in the smart learning environment.

This chapter is organized as follows: Section II introduces the related information required to write this chapter. Section III reviews related work and present methodology. Section IV defines the conceptual model of the smart education ecosystem. Section V presents results in the background of smart learning environments and summarizes the chapter. Finally, Section VI concludes this chapter, results & discussion, and future enhancements.

II. Background Technology

This section presents an overview of digital technology that are commonly clustered into smart education infrastructure.

Over the last decade, many modern technologies have been introduced to the educational sector. These technologies allow educational institutions to do their services in a faster and more efficient way. They enable them to change efforts to capture and keep up the benefits with competitors (Hortense, 2018). These technologies allow them to do better services to their stakeholders and improve communication, collaboration, content management, accessibility to analytics data, and social networking in the educational sector.

Digital Transformation

According to Wikipedia, *digital transformation* is adopting digital technology to transform business services by replacing manual processes with digital processes. It integrates various digital technology into organizations and delivers value to its users/customers. For example, AI-powered chatbots are answering simple customer inquiries in the Web portal. It diminishes the time the customers have to wait to reach a representative. Digital transformation optimizes and makes workflows faster, easier, and more efficient than paperwork business services. The digital transformation includes organizations' processes, models, domain, and culture.

Business organizations innovate, update their business models, and leverages evolving technology. According to Hortense (2018), digital transformation can track metrics and analyze their business services' captured data. Using data-driven insights, business organizations can understand customers' interests, enhance business strategies, and make better decision-making for higher returns. However, digital transformation challenges are employee pushback, lack of expertise, organizational structure, lack of digitization strategy, limited budget, etc. Exponential technological growth began with infrastructure (RedHat, 2020) that determines the kinds of Apps and business processes that worked best. These applications might be monolithic, n-tier, and microservices.

Digital transformation allows human beings and self-directed devices to collaborate, using IT to facilitate modern technologies (e.g., Big Data, AI, Cloud, mobile computing - 3G/4G/5G, and social network technologies - Twitter, Facebook, Instagram, LinkedIn, etc.). In the educational sector, digital transformation requires educational administrators to adopt digital technologies, methods, and approaches. The digital transformation in education can improve performance by allowing innovation and reducing the costs of educational processes. *Digital education* enables students to develop active self-learning skills from anywhere.

Smart Systems

Smart Systems is a new computing technique and architecture that facilitates smart real-world physical systems (Barbor, 2014). It integrate reasoning tasks with sensing, actuation, and communication. It combines many technologies (such as AI, ML/DL/Predictive Analytics, and IoT) with networks, and hence the data become a central part of all application systems. The Smart Systems leverages embedded approaches, networking platforms, and software methods to deliver various services. The Smart Systems core platform combines software systems revolutions and data architectures. It collects, accumulates, incorporates, analyzes, and manages the data from various applications. They solve more complex business complications and lead to the next generation of Smart Systems with significant computing methods and network infrastructure systems.

To design Smart Systems, reliability is a significant concern (Edward, 2018). Probabilistic and deterministic approaches are used to ensure more optimal and practical solutions to Smart Systems. The deterministic model optimizes the Smart Systems environment efficiently and predictably. There is a shortage of data in training data in a smart surroundings in probabilistic approach and limits data-driven strategies.

The IoT technology adopts Smart Systems and gains more prominence in smart environments. The emergence of Multimedia Things results in enormous amounts of streaming data that need to be managed. The use of data-driven techniques can exploit a data-rich ecosystem detailing the smart environment. In Smart Systems, the usual scenarios are the availability of Big Data, the data-driven probabilistic models, and their usage within smart environments. It motivates the traditional rule-based approach to be improved with data-driven strategies supporting optimizations driven by cognitive, ML, and AI that open up novel smart systems design opportunities.

Smart Education

Smart education is an emerging area that makes education more effective by integrating things or objects, technologies, learning environments, pedagogy, etc. It creates innovative approaches, collaboration, and personalized learning that support adaptive, self-motivated, self-regulated, and self-directed learning. *Zhu (2012)* defines "smart education is to create an intelligent environment by using smart technologies so that smart pedagogies can be facilitated as to provide personalized learning services and empower learners, and thus talents of wisdom who have better value orientation, higher thinking quality and stronger conduct ability could be fostered."

Smart education uses advanced technology that helps both students and teachers making themselves for modern education. Educational organizations offer a smart education either physically or virtually (or a blended version of both). It allows smart devices (examples of smartphones, laptops, tablets, etc.) to enhance the traditional education system's teaching and learning process. The advanced learning systems using emerging technologies (such as online classrooms, virtual learning, cloud classrooms, Web learning, etc.) help teachers and students learn extra.

Smart education compromises a model swing in students' mode of access to their learning, i.e., change in education delivery with technologies (Amit, 2018). Figure 1 shows the different roles in the smart education system.

Smart Learning. Smart learning (also known as *digital learning*) provides a complete education to students using recent technology. Teachers and students can adapt the current skills and put them on in their conventional classrooms with emerging technologies. The advantages of smart learning are analytics, reachability, shareability, and transparency.

- *Analytics.* Learning analytics allow smart education to provide continuous improvement and progress against individual learning needs, i.e., *adaptive learning*.
- *Reachability.* Modern learning methods allow wider reachability to students, teachers, and others with the help of technology- for example, virtual learning platforms, online discussion, MOOC, etc.
- *Shareability.* Educational contents can be shared, discussed, and debated now via Microblogging sites or social networking sites, i.e., *social learning*.
- *Transparency.* Smart education opens up a transparent communication channel among stakeholders.

Smart education incorporates human intelligence with tools and smart strategies to bring accuracy to the learning and teaching processes. It is highly beneficial for students with learning disabilities. It encompasses applying a broad spectrum of methods, including virtual learning, blended learning, classroom technologies, e-textbooks, etc. Many educational institutes use hybrid practice in smart education that positively affects student growth and achievement, faculty growth, and effectiveness. Therefore, the educational process' quality grows much better in higher educational institutions (HEI).

Figure 1. Different roles in the smart education system

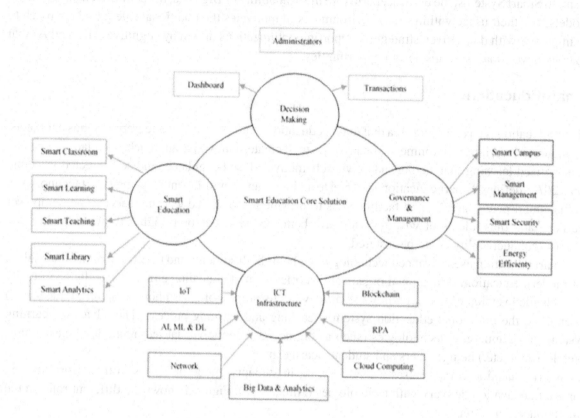

Smart Analytics

Educational institutions face the challenges of managing, analyzing, and visualizing the massive heterogeneous data generated from diverse applications. They make decisions using various practices, approaches, and technologies in the academic field from the generated data. As a result, it builds many theoretical, analytical models. These analytical models may be classified as *data-driven and context-driven*. Academic analytics or educational analytics can be classified as learning analytics, academic analytics, and visual analytics. Both academic analytics and learning analytics are the core of educational data resulted from diverse applications.

- *Academic analytics (AA)* applies the information to manage the academic institutions at the macro level. It focuses on reporting, modeling, analysis, and decision support.
- *Learning analytics (LA)* synthesize massive educational data and predict the outcomes to strengthen insight and impact decision-making through illustrations.
- *Visual analytics (VA)* supports interactive graphical interfaces for information and scientific visualization by combining dissimilar approaches (such as data visualization, data analysis, and insights).

The analytics in educational data play an indisputably important role in improving the quality of education. Quality enhancement could be applied learning process (e.g., learning analytics), educational process (e.g., visual analytics), educational functions, and campus environment services (e.g., academic analytics). IoT technologies improve the quality of education, professional development, and facility management in the education sector.

IoT in Smart Services

Today, IoT becomes the next vital phenomena (Ramakrishnan, 2020) in the educational sector. IoT allows students better access to learning resources (for example, learning materials, communication channels, etc.). With IoT, the teachers can measure the student's learning progress in real-time. *For example*, the automatic attendance system using a smart ID card.

IoT solutions can be used in content delivery, educational business, and stakeholders' healthcare in educational environments. Examples of IoT solutions (Digiteum, 2020) are for educational institutions are EdModo, c-Pen, LocoRobo. Magicard, Kajeet, etc. Some of IoT's critical applications (Infotech, 2020) in the education industry are smart attendance tracking systems, smartboard usage, increased security, addressing disability needs, *sensors for temperature,* and hand-held devices. The IoT provides solutions

Figure 2. Smart IoT platform to Educational institutions

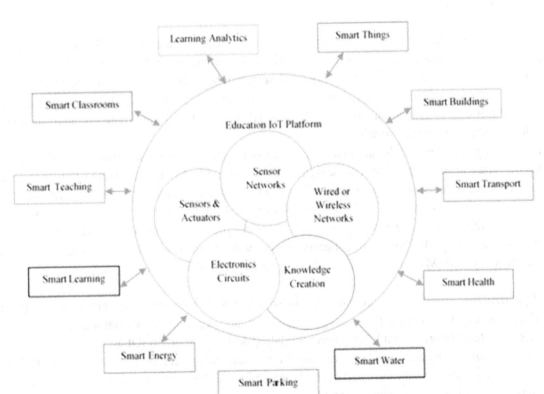

based on integrating information technology to store, retrieve, and process data among individuals or groups.

International Telecommunications Union (ITU) on standards for next-generation networks (NGN) and future networks Bauer (2012) and the IERC (Barnaghi, 2012) provided the definitions of IoT. The IoT creates use of collaborations generated by the convergence of end-user, business methods, and industrial advancement. This convergence can leverage cloud technology to connect smart objects that collect and transmit the data. Some of the IoT enabling technologies are sensor networks, Radio Frequency Identification (RFID), machine-to-machine (M2M), WSN (wireless sensor network), mobile computing technologies, semantic data and search, IPv6, etc., and they provide warnings or alerts, security, energy consumption, process automation, and performing in a single ecosystem.

Traditional products and services such as bots, smart buildings, smart health services, smart vehicles are a few examples of IoT ecosystems. The IoT enables smart services (Minghui, 2020) in various fields like smart healthcare, smart logistics, smart transportation, smart manufacturing, smart education, etc. These products can provide energy and efficiency savings and generally deliver on investment.

An illustration of a Smart IoT platform for educational institutions is presented in Figure 2.

In IoT systems, data management is a severe issue in the environment. The growing number of IoT devices creates an enormous capacity of unstructured data in real-time data that is critical (Ovidiu, 2013). Many technologies are involved in data management for data collection, process, analytics, and complex event processing. It is mandatory to develop a data ecosystem for creating sustainable architectures and models.

Data Ecosystem

The digital transformation creates a data ecosystem for smart systems that need significant data-rich smart environments. Educational organizations create their ecosystem model and architecture to gather, store, investigate the data, and act upon it. The data ecosystem can be built with a product analytics platform for analytics. These data analytics platform helps data scientist and analysts to integrate multiple data sources. Educational organizations use AI, ML, and DL approaches to automate the analysis process and calculate the performance metrics.

A data ecosystem deals with data management in smart learning environments. From the *usability perspective*, the learning data ecosystem design needs to be simple. A data ecosystem solution may need several architectural decisions as far as storage is concerned. Other architectural selections can be related to high storage, speed, reliability, and data security. From an *architectural perspective*, the devices need to be aligned with the learning environment. These devices must be synchronized and enabled through a well-architected, well-designed, and coherent process addressing all requirements. The fundamentals of the data ecosystem (MixPanel, 2020) are infrastructure, analytics, and applications.

1. *Infrastructure* provides the hardware and software services that include servers, search languages (like SQL), and hosting platforms to capture, collect, and organize the educational data. It may use Big Data technologies for a large volume of data (like Hadoop or NoSQL).
2. *Analytics platforms* search and summarize all data available in a central place. They can dig deeper into the data, offer an interface, and make designs more quickly. They analyze all the specific user's data, their actions, and predict the results.
3. *Applications* accept the data, process, and make them usable to the stakeholders.

According to Curry (2018), many data ecosystems exist in a smart environment. They are directed, acknowledged, collaborative, and virtual. *Directed data ecosystems* can be centrally organized to accomplish a precise determination of educational organizations. *Acknowledged data ecosystems* have well-defined concepts and shared committed resources for collaboration among the members. *Collaborative data ecosystems* allow members to interact for central purpose and decide to enforce and sustain standards. *Virtual data ecosystems* allow members to share decentralized resources to realize explicit goals.

Modern data ecosystem using Cloud and AI-driven enable an educational organization to generate maximum business value from its unique data assets. The modern data ecosystem enables (Curry, 2018) many challenges to the smart environment concerning infrastructure, governance, systems engineering, and human-centricity.

For example, an IoT Ecosystem (Mehmet Yildiz, 2017) combines various tiers from the user to the connectivity. In a typical IoT ecosystem (Mary Allen, 2018), the components are connected to the computing platforms or cloud through the Internet. The IoT ecosystem has many enabling technologies for connecting, controlling, and deriving value from their connected things in elegant atmospheres. Also, Cloud Computing facilitates data integration effectively. The deployment of IoT applications creates smart environments in educational institutions. The advantages of IoT in education are mobility, scalability, and feasibility in a smart environment.

Learning Ecosystem

A learning ecosystem is the collection of all learning sources that learners can access in the educational institutions. It is typically comprised of significant elements - learners, learning resources, and technological infrastructure. The learners are the stakeholders who use the learning resources generated by educational organizations. The technical infrastructure facilitates learning management within an organization (e.g., LMS). A learning ecosystem allows educational organizations to avoid internal stagnation that utilizes LMS. A learning ecosystem creates a seamless learning experience for learners within the educational organization.

- It has a platform, which allows teachers to introduce students via LMS.
- It provides continuing educational experiences.
- It allows the organization to streamline its enhancements.
- It allows stakeholders to work with one another simultaneously.

Educational Institutions can create practical and streamlined learning ecosystems.

III. LITERATURE REVIEW AND METHODOLOGY

In the circumstance of a smart learning environment (SLE), the design needs to extract useful and accurate insights from the data generated by an intelligent learning environment to create significant conclusions for the business in education. Hence, the literature review included smart education, learning data ecosystem, data management, and IoT.

Jason Ng et al. (2014) introduced peer-to-peer learning to facilitate the modern landscape that bears formal and informal learning atmospheres. The *quality of education* is significantly increased the qualified teachers and make them more efficient (Mrinmoy, 2020) (e.g., automated grading, automated attendance, biometric authentication, chatbots, text translation systems, and personalization). *Archie (2016)* framed the work and made improvements to the collection of data in higher educations. Mrinmoy (2020) stated that AI and ML models could be used to predict students' studies, drop out, and develop appropriate remedial measures.

Demchenko (2014) discussed paradigm change from traditional architecture to data-centric architecture and operational models in the big data ecosystem. *Hui Deng (2016)* introduced a sustainable big data ecosystem model. *Kitsios (2017)* examined the business model required in an open data-based business system. *Csaba Csáki (2019)* prepared a data ecosystem model for visual notation. *Oliveira (2019)* studied the current literature on data ecosystems.

According to *Ray (2016)*, the lack of architectural knowledge resisted the researchers to get through the IoT-centric approaches. The IoT-oriented architectures enable the distribution of lower-cost sensors, adopt intelligent systems and gain more data visibility into smart environments (Curry 2018). *Shafique (2020)* presented the IoT technology with an extensive overview of the emerging 5G-IoT scenario.

Lee (2020) examined the smart education model - a four-tier framework of smart pedagogies and the SLE's critical features. *Hoel (2018)* highlighted a cognitive smart learning model and a smartness level model for SLE. *Putra (2019)* created an educational service system model to improve the quality of high-school education services in West Java. *Ji-Seong (2013)* introduced a cloud-based smart education system for e-Learning content services to deliver and share various enhanced forms of educational content.

Sakshi (2017) proposed a business model of smart education based on the IoT that enables smart objects and provide communication within the smart environment. *Al-Majeed (2014)* proposed a smart education environment system framework to meet an emerging IT-aware generation's interests. According to *Marc* (2016), the IoT's impact on next-generation smart environments would depend on integrating IoT and cloud computing in next-generation smart environments. *Minghui (2020)* investigated an edge-driven security framework for intelligent IoT systems.

Oliveira (2019) stated that data ecosystems were socio-technical complex networks and collaborated data to support new businesses. *Edward Curry (2020)* explored a data ecosystem's role in designing intelligent systems to support data-rich IoT-based smart environments. *Lior Shalom (2020)* developed a Robot Operating System (ROS), a framework to develop, build, and deploy software for robotics. ROS creates a valuable ecosystem with extensive documentation, utilities, and tools.

UDNP (2016) redressed a series of recommendations for diverse communities of data stakeholders. Some of the suggestions are as follows:

1. It accomplishes several inputs from the data ecosystem to ensure the quality of data.
2. It ensures that data users and other stakeholders must know how to access and use the data.
3. It ensures the availability of and access to data to the stakeholders.
4. It adopts data standards, including metadata, sharing, and interoperability.
5. It incorporates data innovations and big data approaches into monitoring and governance.

Findings

The SLE generates heterogeneous data in educational organizations. The increasing number of devices in SLE can produce massive amounts of educational data. As a result, the learning data ecosystem consisting of structured (SQL) or unstructured data (e.g., pictures, video, audio, and documents) and semistructured data detailing the SLE that can be demoralized by data-driven methods. The conventional approaches such as AI techniques, machine learning, deep learning, cognitive, and learning analytics, which support optimizations, open new possibilities for designing data ecosystems to the educational environment.

Build a Data Ecosystem

To build a data ecosystem, the designers are frustrated and struggled to integrate data within the organization. It has limitations in its flexibility to adapt to changing experiences, people, relationships, partnerships, the workforce, organizations, and the environment. For scalability, the educational organizations can modernize their data architectures by introducing Data Lake technology against the traditional Data Warehouse. A data lake is centralized data management, creating a next-generation data platform enabling self-service data access for their analysts. The Data Lake technology makes a next-generation data architecture to educational organizations, i.e., *data-driven*. The next-generation data architecture helps educational institutions to progress more collaboration, generate additional revenue, digital transformation in educational processes to increase efficiency.

Building a next-generation data ecosystem for educational organizations enables them to share their data with diverse applications and ingest data from external data sources to create innovative data solutions quickly. This solution can be integrated with the core solution to drive scale, enable fast delivery, and provide flexibility to help educational institutions grow and increase customer value. For insights, AI, ML, and DL algorithms create predictive and prescriptive analytics and deliver insights-as-a-service.

BI tools provide analytics visualization and omnichannel interaction among core solutions with diverse applications. With AI, ML, and DL-driven BI solutions, educational institutions can optimize educational processes.

Methodology

This chapter uses a mixed-method qualitative research methodology that uses the analysis of the existing status, difficulties, and smart learning requirements. This research is carried out in 4 phases. *Phase 1* explores the present status of complications and the need for a next-generation smart education ecosystem. *Phase 2* studies the best procedures and approaches in the data ecosystem. *Phase 3* formulate the conceptual model and *Phase 4* evaluates the proposed model with quality parameters.

IV. RESULTS AND DISCUSSIONS

In the digital age, smart education used live chat, educational data centers, video surveillance, video conferencing, and location-based services. These services create large volumes of mixed data such as complex, unstructured text, animated gifs, audio, and video data. To handle the mixed and operations, educational institutions require a scalable learning data ecosystem. The learning data ecosystem enables the process and storage of large volumes of data (i.e., structured, semistructured, and unstructured) generated by internal and external systems - this discussion about the design of next-generation education data ecosystem for modern educational institutions.

Next-Generation Education Data Ecosystem

Educational organizations must evolve their data and information architecture to establish a next-gen data ecosystem successfully. It enables them to discover business insights and drive quick responsive business decisions—the next-generation data ecosystem support both Data Warehouse and the Big Data features. Data analytics, data arrangements, data ingestion, data modeling, data quality, data type, and data volume are data analytics. Figure 3 shows the next-generation smart education data ecosystem.

Figure 3. Next-generation smart education data ecosystem

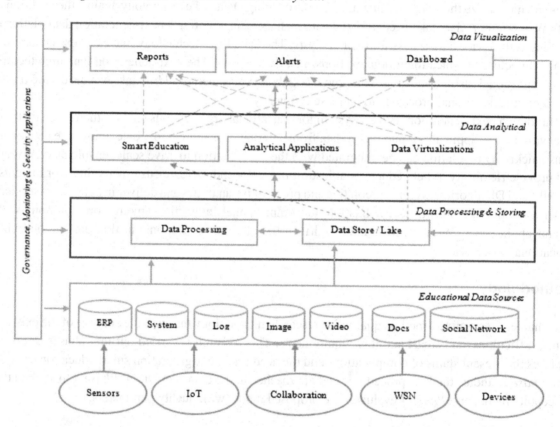

- *Data Analytics*. It supports all types of analytics.
- *Data Arrangements*. It supports consolidated, integrated, aggregated, and virtualized data.
- *Data Ingestion*. It supports both batch and stream processes.
- *Data Modeling*. It supports various models and schema as per end-user requirements.
- *Data Quality*. It supports all levels of data quality for structured and unstructured.
- *Data Type*. It supports structured, unstructured, and multi-structured.
- *Data Volume*. It supports from terabytes to exabytes and beyond.

The next-generation data ecosystem serves as the core data tier for educational institutions that form the data lake. This data lake is populated with different data types from diverse sources in a scalable data storage for analytical applications.

Design Challenges

The design challenges of an education data ecosystem are data integration and data linking. *Data integration* focuses on integrating social network sites, sensors, and IoT systems in a personalized, inflexible, and reusable style. *Data Linking* allows the created data to be shared and reused by many educational organizations. According to *Alexandru (2018)*, the design principles considered in the design of the data ecosystem is listed and presented below:

- *Flexibility* is quickly reformed varied set of components and services.
- *Heterogeneity* must consider various entities, patterns, standards, and types of hardware and software platforms.
- *Scalability* supports many connected entities, devices in a flexible way.
- *Security* include multilevel security procedures, containing identity, authorization, authentication, and data protection.
- *Weak Coupling* allows processing modules, communication medium, and digital entities may be decoupled in processing time and storage space.

Design Factors and Characteristics

Building a data ecosystem can add more data for easy access to them. When creating a data ecosystem for SLE, the things to be considered are a data analytics and data governance. Educational institutions integrate *data analytics* platforms with their business operations that allow every stakeholder to access data. *Data governance* may include data capture, data usage and storage, data safeguard, and disposal.

Several design characteristics are considered in data ecosystems for the operation, depending on the circumstances. They are infrastructure, availability, privacy, formats, and services. It has many technical characteristics: infrastructure, data availability, data privacy, data formats, and data services. These characteristics disturb the infrastructure that supports data distribution within the educational data ecosystem.

Educational data ecosystems support consistent and synchronized data movements, seamlessly move data between many systems. Hence, the educational data ecosystem supports principles such as searchability, accessibility, interoperability, and reusability. These principles can impact processes and workflows for educational data. They improve the data infrastructure and reusability and lead to an academic data ecosystem supporting transparency, reproducibility, and reusability.

Requirements

It is performed a systematic analysis to identify the functional and non-functional requirements to the educational system. These requirements can be data process, data links, query, security, and management, as shown in Table 1.

Table 1.

S/N	Requirements	Importance
1	Data Discoverability and Accessibility	Medium
2	Data Linkage	Medium
3	Data Management	Medium
4	Query on real-time data	High
5	Query on historical data	High
6	Real-time data	High
7	Security	Medium

Functional View

The next-generation educational data ecosystem entails a systematic assessment of educational data and multi-stakeholder problems in the academic environment. It requires an evaluation infrastructure, which involves data gathering, distribution, and data use and taking advantage of new technologies for participation. This data value chain shown in Figure 4 ensures the continued scalability and enhancement of educational institutions' statistical capacities.

Figure 4. The data value chain

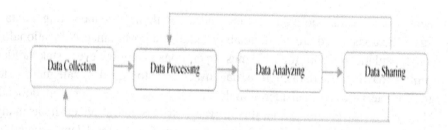

Educational institutions create a next-generation educational data ecosystem for capturing and analyzing data paths to make valuable insights. The data environments in the next-generation educational data ecosystem are designed to grow. The next-generation educational data ecosystem's functional view comprises many components such as interaction, analytics, application, integration, and user interface. The significant elements in the next-generation educational data ecosystem are infrastructure, analytics, and applications.

- *Infrastructure.* The infrastructure can be hardware (servers, network, virtualization, storage, etc.) and software (SQL), and accommodating platforms. Infrastructure catch and store different types of data - organized, un organized, and semi- organized. The infrastructure holds a large volume of data.

- *Analytics.* Analytics platforms have stakeholders and analytical tools. They help stakeholders integrate multiple data sources from numerous educational applications. They provide ML and DL tools to automate the analytical process. They identify stakeholders and their data, process and anticipate their next actions.
- *Applications.* Applications are services that act on the captured data to make it usable.

The next-generation educational data ecosystem should be innovative, integrated, infrastructure, and support educational transactions. The innovation provides a foundation on which to develop complementary services. The integration can be combined with an innovation platform. The infrastructure support data sharing and reuse for data ecosystem for smart learning environments. The communication approaches used in the data ecosystem and is publisher-subscriber, request-response, and action-based communication. For example, the publisher node posts messages into a channel, other nodes can subscribe to this channel. Here, the communication can be is one-way.

ARCHITECTURE VIEW

The next-generation educational data ecosystem offers appropriate channels for delivering new applications and developing new business models. It requires an innovative architecture that supports highly complex and inter-connected educational applications to enable various emerging technologies. It requires platforms to effectively abstract actionable information from the massive raw data that keeps the education system's real-time and synchronization. The proposed next-generation educational data ecosystem's salient features require educational institutions to institute data governance, metadata, data interoperability, data quality, and data security.

- *Data Governance* identifies the roles of Big Data initiatives to educational organizations.
- *Data Quality* set up processes to enhance the quality of unstructured data coming from unconventional sources.
- *Data Security* integrates sufficient data authentication, authorization, and encryption capabilities across the next-generation data ecosystem.
- *Metadata* ensures data interoperability and defines the context of the structured and unstructured data in the data lake environment.

Educational institutions must evolve their data and information architecture that successfully establish a next-generation data ecosystem. This architecture enables them to discover business insights and drive quick and responsive business decisions. Figure 5 depicts the next-generation data ecosystem's architecture that encapsulates the characteristics of both the traditional data (Data Warehouse and data mining technique) and the current data (Big Data) ecosystem.

In the proposed next-generation data ecosystem, a big data platform serves as the core data, i.e., the data lake. This data lake is occupied with different types of data from diverse sources. In the data ingestion, data is ingested into the core data using a combination of batch or real-time processing. Business analytical applications and educational systems can consume data from the data lake with various services such as data-as-a-service, analytical service, query services, learning content services, storage services, and virtualization.

Figure 5. The proposed architecture of the next-generation data ecosystem

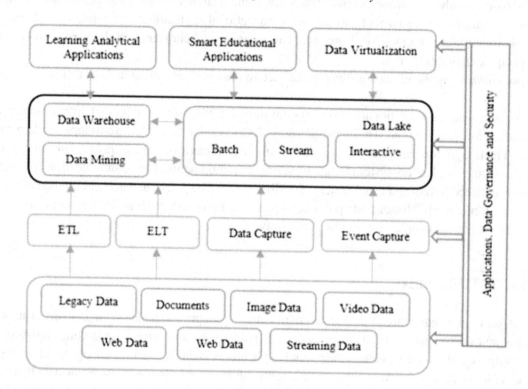

Conceptual Model View

Building a robust next-generation educational data ecosystem is a responsive task and agile process. The data analysts must be allocated accountabilities allowing the skills showed by both. The data must be ingested from diverse sources, stores in a different format, and then be analyzed before the final presentation. These lengthy processes can take weeks, months, or years. The proposed conceptual model consolidates the data visualization tier, data analytics tier, data processing tier, and the data source tier, as presented in Figure 6.

Data Source Tier. The data source tier (or ingestion tier) offers a data source service to the data ecosystem. The raw data comes from many sources, including internal sources, external sources, relational databases, and non-relational databases, i.e., from social media, smartphones, e-mails, log files, etc. The data source tier pulls the raw data from various applications installed in the SLE. The central data sources are smart education services, ERP systems, sensors, IoT devices, social networks, blogs, video conferencing, users' internet history, Weblogs, etc.

A data ecosystem infrastructure for educational institutions can be used to collect and accumulate dissimilar data, i.e., structured, unstructured, and multi-structured. The data source collects data from smart learning environments through various educational applications and sensors. The unstructured and semistructured data should be transformed into a structured format for further processing. The *structured data* is cleaned, labeled, and organized, e.g., ERP, LMS, management systems, etc. The *unstructured data* has not been scheduled for analysis, e.g., students browsing data, social network data, log data, sensor data, IoT data, Video data, etc. The semistructured (or multi-structured) data could be a mixture of both.

Figure 6. A conceptual model to the smart learning data ecosystem

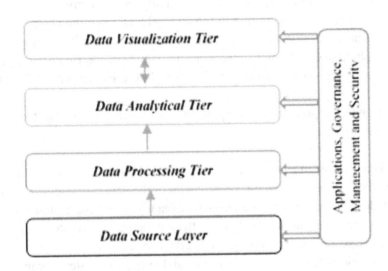

The data ingestion may be in batch, interactive, and streaming.

1. *Batch processing* collects large groups of data and delivers them together. The data collection may happen because of conditions, such as an ad-hoc, or maybe based on a schedule- examples of sensors and IoT data.
2. *Interactive processing* accepts input from humans in the form of data and commands. The response can be problematic if the system is conducting social interaction systems to achieve through social interfaces.
3. *Stream processing* is a regular flow of data and vital for real-time analytics. It requires comparatively more resources because it is continuously monitoring any changes in the data pools. Examples of video surveillance, video conferencing, etc.

The RFID and the WSN identify and acquire learning data efficiently and reliably in an SLE. They provide well-organized application services for IoT devices and sensors installed in SLE. The data source tier ensures the quality of data. The corrupt data may not result in quality insights.

Data Processing Tier. The data processing contains of a huge number of edge sensors and devices with data pre-processing and storing services. They can provide communication services and contents to IoT sensors effectively with low communication delay. The data pre-processing is accomplished in the data processing tier that offers powerful computing and storage capacities to support various services. The analytics provides high scalability, flexibility, and availability to applications for devices and sensors. The data processing module takes together all the previous data segments and passes them to the tools for shaping them into insights. The different types of analytics are descriptive, diagnostic, predictive, and prescriptive.

1. *Descriptive* analytics described the current standing of a business based on old data. It uses the previous data sequence patterns to forecast seasonal impacts, marketing, sales, etc. These data and consumer insights help in making sense of the internal metrics of the business.
2. *Diagnostic*. This analytics dive deeper into the data and explains why the actions did not produce the results. It explains in detail which of the measures could not contribute to the projection metrics.
3. *Predictive* analytics project the future outcome based on previous data. It highlights trends and evaluates metrics, and predicts future results.
4. *Prescriptive* analytics is putting in the inputs and actions in the system. It shows businesses the best way to move forward.

The processed data is stored in a Data Lake or a Data Warehouse and then eventually processed. The data warehouse and data lake are an essential component of the data ecosystem. They must be efficient and relevant to provide quick processing and needs to be readily accessible. Data Lakes store the original data, and they can be used later on. In warehouses, the data are grouped into categories and stored. A data lake involves a massive amount of storage than a data warehouse.

Downstream applications (e.g., business analytics and enterprise systems) can consume data from the data lake using various services such as data-as-a-service (DaaS), analytical query services, content services, and spatial services virtualization. The workflow and scheduler applications specify the sequence for the execution of tasks based on their dependencies on completing other jobs.

Data Analysis Tier. The data analysis is the core module of the educational data ecosystem. Data analytics provides various analytical services for data visualization. The data analytics supports the analytical platform to develop, run, and operate the applications. The data analytics provide the required software services through the network/Internet. The analytical applications lead to a new digital framework for producing many novel applications and services. The AI and ML technologies play a vital role by assisting service providers to offer IoT devices with varied and personalized services for data analysis and data prediction in the application.

Data visualization Tier. Data visualization (or consumption) involves the presentation of the insights to the analytical user. The data consumption can be in the form of alerts, notifications, tables, graphs, visualizations, etc. The stakeholders are students, teachers, researchers, recruiters, parents, evaluation specialists, governing bodies, content creators, and decision-makers. The most crucial point is that insights should be precise and understandable. Educational institutions can develop a new policy and make changes in their business operations.

Advantages and Challenges

The proposed data ecosystem's main advantages are flexibility, location awareness, low latency, operational cost, high reliability, and scalability.

-   *Flexibility*. It can be achieved using virtualization that shares the heterogeneous infrastructure and multiple services in the SLE.
-   *Location of Awareness*. The location of stakeholders can provide personalized, contextual services and mobility.
-   *Low Latency*. Smart education services can be placed close to the stakeholders to support real-time services.

- *Operational cost.* Integrating various services can reduce operating costs in the educational environment.
- *Reliability.* Integrating educational services with a cloud environment can provide reliable service to the stakeholders.
- *Scalability.* It supports scalability to computing, data storage, connected devices, and services.

Moreover, education administrators need to overcome some of the issues before sensors and IoT can be fully installed in the educational environment. These devices are high costs, class ethics, data processing infrastructure, and security.

Evaluation

The recommended model has been evaluated using data volume, data ingestion, data types, data process, data model, data quality, data analytics, data arrangement, etc., as shown in Table 2.

Table 2. Features considered for evaluation

S/N	Features	Supports
1	data analytics	It supports prescriptive and predictive analytics.
2	data arrangement	It supports consolidated, integrated, and aggregated.
3	data ingestion	It supports both ETL and ELT.
4	data model	It supports the model as per requirements.
5	data process	It supports both batch and streaming processes.
6	data quality	It supports quality for both structured and unstructured.
7	data types	It supports structured, unstructured, and multi-structured.
8	data volume	It supports both data and video analytics

The educational data ecosystem supports a more classified and end-to-end method to data creation, usage, investigation, and communication. The data deluge is the well-known valuation of the data ecosystems. The datasets' design was specified for the nonexistence of adherence to standards, information sharing, and coordination between data producing agencies. Introducing standards for data collection, facilitating data literacy, ensuring the availability, and accessibility to data, data aggregation, ensuring transparency and open data can improve the ICT infrastructure in the monitoring frameworks. All data stakeholders can ensure collaborative partnerships and fostering innovations in educational organizations.

V. FUTURE RESEARCH DIRECTION

In the educational sector, technology plays a critical role in digital transformation. Technology allows an integrated and coordinated action by the stakeholders for sustainable development. In the smart education area, some research opportunities, which are identified, include data connectivity, network connectivity, flexibility, and scalability of the advanced systems.

1. Smart security, a research opportunity as smart education works with the personalized data of stakeholders.
2. Integrating with blockchain technology provides more security features in the smart education system.
3. Research on recommendation system focuses on detecting many problems and provide suggestions and recommendations to the stakeholders.
4. Research on data visualization techniques can improve the teachers and students' exploiting the new information given to them.

In addition to the above, the smart education data ecosystem requires many challenges to overcome regarding learning infrastructure, data engineering, administration, and human-centricity. Research is needed on a data platform, distributed data authority models for data ecosystems, and trust in algorithms and data to make data-driven critical decisions.

VI. CONCLUSION

Digital transformation motivates a data-rich smart environments with massive data, which results in ecosystems that present challenges and chances in smart systems design. The structure of data ecosystems is becoming an important topic nowadays. Data ecosystems offer many options; however, it introduces many challenges for researchers and developers. The IoT is a new technology that connects device-to-device, device to people, a device to place, etc. In a smart education environment, IoT technology generates tremendous amounts of data. This vast data or big data can then create valuable new services in the learning environment.

This chapter introduces the technology-neutral model for the digital data ecosystem to a smart education system. The proposed model consists of data presentation, data analytics, data storage, and data source tiers.

REFERENCES

Allen, M. (2018). *Building the IoT ecosystem.* https://insightaas.com/building-the-iot-ecosystem/

Averian, A. (2018). *A Reference Architecture for Digital Ecosystems, edited by Jaydip Sen, Internet of Things Technology.* In *Applications and Standardization.* IntechOpen. doi:10.5772/intechopen.70907

Barcelo, M. (2016). IoT-Cloud Service Optimization in Next-Generation Smart Environments. Academic Press.

Barnaghi, P., Wang, W., Henson, C., & Taylor, K. (2012). Semantics for the Internet of Things: Early Progress and Back to the Future. *Intel Journal on Semantic Web and Information Systems, 8*(1), 1–21. doi:10.4018/jswis.2012010101

Bauer, M. (2012). *Deliverable D1.4 - Converged architectural reference model for the IoT V2.0.* http://www.iot-a.eu/public/publicdocuments/documents-1/1/1/D1.4/at_download/file

Boulton, C. (2020). *What is digital transformation? A necessary disruption.* https://www.cio.com/article/3211428/what-is-digital-transformation-a-necessary-disruption.html

Csáki, C. (2019). *Open Data Ecosystems: A Comparison of Visual Models. In Electronic Government and the Information Systems Perspective.* Springer. doi:10.1007/978-3-030-27523-5_2

Cubarrubia, A., & Perry, P. (2016). *Creating a Thriving Postsecondary Education Data Ecosystem.* http://www.ihep.org/research/publications/creating-thriving-postsecondary-education-data-ecosystem

Curry, E., & Ojo, A. (2020). Enabling Knowledge Flows in an Intelligent Systems Data Ecosystem, Real-time Linked Dataspaces. doi:10.1007/978-3-030-29665-0_2

Curry, E., & Sheth, A. (2018). Next-Generation Smart Environments: From System of Systems to Data Ecosystems. *IEEE Intelligent Systems, 33*(3), 69–76. doi:10.1109/MIS.2018.033001418

Curry, E., & Sheth, A. (2018). Next-Generation Smart Environments: From System of Systems to Data Ecosystems. *IEEE Intelligent Systems,* 69–75.

Dai, M., Su, Z., Li, R., Wang, Y., Ni, J., & Fang, D. (2020). An Edge-Driven Security Framework for Intelligent Internet of Things. *IEEE Network, 34*(5), 39–45. Advance online publication. doi:10.1109/MNET.011.2000068

de la Boutetière, H., Montagner, A., & Reich, A. (2018). *Unlocking success in digital transformations.* https://www.mckinsey.com/business-functions/organization/our-insights/unlocking-success-in-digital-transformations

Demchenko, Y., Laat de, C., & Membrey, P. (2014). Defining architecture components of the Big Data Ecosystem. *Intel. Conf. on Collaboration Technologies and Systems (CTS),* 104-112. . doi:10.1109/CTS.2014.6867550

Deng, H. (2016). Big data ecosystem model, and application in the city. *Journal of Big Data Research, 2*(2), 68–75.

Digital Transformation. (n.d.). In *Wikipedia.* https://en.wikipedia.org/wiki/Digital_transformation

Digiteum. (2020). *How IoT is used in Education: IoT Applications in Education.* https://www.digiteum.com/iot-applications-education

Dua, A. (2018). *Smart education is more than just Advanced Learning Methods.* https://yourstory.com/2018/05/smart-education-advanced-learning

Frost, S. (n.d.). *Mega Trends: Smart is the New Green.* https://www.frost.com/prod/servlet/our-services-age.pag?mode=open&sid=230169625

Hannon, V., Patton, A., & Temperley, J. (2011). *Developing an Innovation Ecosystem for Education.* White Paper, Cisco Innovation Unit.

Hoel, T., & Mason, J. (2018). Standards for smart education – towards a development framework. *Smart Learn. Environ., 5*(1), 3. doi:10.118640561-018-0052-3

I-Scoop. (2020). *Digital transformation: online guide to digital business transformation*. https://www.i-scoop.eu/digital-transformation/

Infotech. (2020). *Smart Classroom Technology-IoT in Education Industry*. https://medium.com/@chapter247infotech

Jason, N. (2014). Smart Learning for the Next Generation Education Environment. In *2014 International Conference on Intelligent Environments*. IEEE Xplore. 10.1109/IE.2014.73

Jeong, J.-S., Kim, M., & Yoo, K.-H. (2013). A Content Oriented Smart Education System based on Cloud Computing. *Intel. J of Multimedia and Ubiquitous Engineering*, *8*(6), 313–328. doi:10.14257/ijmue.2013.8.6.31

Kitsios, F., Papachristos, N., & Kamariotou, M. (2017). Business Models for Open Data Ecosystem: Challenges and Motivations for Entrepreneurship and Innovation. *IEEE 19th Conf. on Business Informatics (CBI)*, 398-407. 10.1109/CBI.2017.51

Lee, R. S. T. (2020). Smart Education. In *Artificial Intelligence in Daily Life*. Springer. doi:10.1007/978-981-15-7695-9_11

MixPanel. (2020). *How to create a Data Ecosystem*. https://mixpanel.com/topics/what-is-a-data-ecosystem/

Oliveira, S., Barros Lima, G. D. F., & Farias Lóscio, B. (2019). Investigations into Data Ecosystems: A systematic mapping study. *Knowledge and Information Systems*, *61*, 589–630. doi:10.100710115-018-1323-6

Oriel, A. (2020). *Building a Strong Data Ecosystem for AI-Powered Organizations*. https://www.analyticsinsight.net/building-strong-data-ecosystem-ai-powered-organizations/

Parthasarathy, S., Tung, T., Munnelly, S., & Joshi, S. K. (2019). *Bringing Data Together: A Modern Data Ecosystem*. Accenture. https://www.accenture.com

Putra, R. R. J., & Putro, B. L. (2019). *J. Phys.: Conf. Ser. 1280 032029*. IOP Publishing. doi:10.1088/1742-6596/1280/3/032029

Raman, R. (2020). *IoT and its impact on education*. https://www.deccanherald.com/supplements/dh-education/iot-and-its-impact-on-education-845493.html

Ray P P (2016). A survey on Internet of Things architectures. *Journal of King Saud University – Computer and Information Sciences*, 1319-1578. . doi:10.1016/j.jksuci.2016.10.003

RedHat. (2020). *What is digital transformation?* https://www.redhat.com/en/topics/digital-transformation/what-is-digital-transformation

Roy, M. (2020). AI Intervention in Education Systems of India: An Analysis. *Solid State Technology*, *63*(2), 1395–1402.

Shafique, K., Khawaja, B. A., Sabir, F., Qazi, S., & Mustaqim, M. (2020). Internet of Things (IoT) for Next-Generation Smart Systems: A Review of Current Challenges, Future Trends and Prospects for Emerging 5G-IoT Scenarios. IEEE Access, 8, 23022-23040. doi:10.1109/ACCESS.2020.2970118

Smart Systems and Internet of Things Platforms. (n.d.). *Overview of Research and Analysis and Summary Findings, Smart System Design*. Barbor Research. https://niolabs.com/app/uploads/2017/10/HRI_Platform-Rpt-Summary_12-October-2017.pdf

Subramanian & Srivastava. (2017). *Architecture Patterns for the Next-generation Data Ecosystem*. Tata Consulting Services.

Tassey, M., Gray, E., & Cottrell, S. (2020). *Data Transfer in the Larger Education Ecosystem*. United States Department of Education, Privacy Technical Assistance Center. https://studentprivacy.ed.gov/sites/default/files/resource_document/file/DataTransfer-in-the-Larger-Education-Ecosystem.pdf

UNDP. (2016). *Data Ecosystems for Sustainable Development, an Assessment of Six Pilot Countries*. Report, United Nations Development Programme.

Vaitsis, C., Hervatis, V., & Zary, N. (2016). Introduction to Big Data in Education and Its Contribution to the Quality Improvement Processes. In. Big Data on Real-World Applications. IntechOpen Science. doi:10.5772/63896

Vermesan, O., & Friess, P. (2013). *Internet of Things: Converging Technologies for Smart Environments and Integrated Ecosystems*. River Publishers Series in Communications.

Yildiz, M. (2017). *Introduction to IoT Ecosystem, a Technical, Architectural & Solution*. https://medium.com/illumination-curated/introduction-to-iot-ecosystem-25b359c8cf23

Chapter 7
Smart IoT Systems:
Data Analytics, Secure Smart Home, and Challenges

Ritu Chauhan
Amity University, India

Sandhya Avasthi
(iD) https://orcid.org/0000-0003-3828-0813
Amity University, India

Bhavya Alankar
Jamia Hamdard, India

Harleen Kaur
Jamia Hamdard, India

ABSTRACT

The IoT or the internet of things started as a technology to connect everyday objects over the internet, which has evolved into something big and invaded into every single aspect of our lives. As technology is gaining momentum, IoT-based smart devices usage among users is expanding, which generates massive data at our disposal across various domains. The authors have systematically studied the taxonomy of data analytics and the benefits of using advanced machine learning techniques in converting data into valuable assets. In the studies, they have identified and did due diligence on different smart home systems, their features, and configuration. During this course of study, they have also identified the vulnerability of such a system and threats associated with these vulnerabilities in a secure smart home environment.

DOI: 10.4018/978-1-7998-7541-3.ch007

1 INTRODUCTION

Recent advancements in the internet of things and improvements in computing devices have made communication between devices easy. It is fast becoming a topic of social, technical, and economic significance. A wide range of consumer goods, mobile devices, cars, industrial components, utility items, sensors, and other objects are combined with internet connectivity and powerful data analytics capabilities to transform the life around us. The forecast says that by 2030 total connections to the internet will be more than 50 billion and globally impacting more than $11 trillion. McKinsey stated a 300 percent rise in linked IoT devices in 2013 annual report and valued the future economic effect of IoT at $2.7 trillion to $6.2 trillion annually by 2025 (Manyika et al., 2013; Manyika et al., 2015). Because users are now very dependent on different applications over the internet, technologies like the Internet of Things (IoT) will be more widespread in coming years. The embedded technology that includes wired and wireless communication along with sensors, actuators, and physical devices is known as IoT. The aim of smart computing devices is augmentation and effortlessness in the experience provided to the users. Hence, IoT is fast becoming a key source of new data, that demands storage and analysis needs. The IoT offers endless ways to connect everyday objects which a common person uses inside and outside the home. In terms of innovation, this area is wide open accelerating demands for machine learning and other interconnecting technologies. It may be an encouraging time for creative people, partially because understanding these interconnections is still arduous. The system offers both prospects and potential security problems.

Smart devices like smart phone, vehicles, temperature control systems, smart elevators, healthcare devices and automation systems are making our life easier and better. Large number of IoT devices being used and their support systems generates massive amount of data that needs storage and processing at cloud storage centers. The massive data generated in the process cannot used for knowledge discovery unless processed by advanced machine learning techniques that can handle it properly. The IoT application domain range from social media, smart healthcare, smart e-agriculture, smart electricity, and smart vehicles. Moreover, the IoT success depends on specific protocols to interconnect such as application-layer protocols that interacts with users directly. The protocols such as Constrained Application Protocol (CoAP), Hyper Text Transfer Protocol (HTTP), Message Queuing Telemetry Transfer (MQTT), and Advanced Message Queuing Protocol (AMQP) (Karagiannis et al., 2015; Zschörnig et al., 2020) are responsible for communication between devices.

Intelligent living occurs, as smart devices, apps and utilities work together to create an environment that surrounds us. The IoT based devices help people live their lives comfortably and securely, so that they can concentrate on what really matters. The Statista (Forecast end-user spending on IoT solutions worldwide from 2017 to 2025, n.d.) forecast predicts that global demand for smart devices is growing, with shipments to surpass three billion by 2023 for devices such as smartphones, PCs, laptops, wearable bands, smart speakers, and smart personal audio systems. The demand for smart speakers is expected to grow with shipments set to exceed 39 million units in 2023, because speakers can serve as command centers in smart homes.

The contribution of the chapter are as follows:

- review different machine learning application for IoT data.
- Taxonomy of machine learning algorithms.
- The features of IoT data in real-time.
- Smart city as the application of IoT.

- Challenges in secure smart home systems and IoT data analytics.

We studied IoT concepts, applications, and importance of data analytics on IoT data in this chapter. We discuss in detail the different IoT paradigm and their integration with data analytics methods. In section I we present an introduction to Internet of Things and latest trends. In the section II and III, the IoT components and machine learning integration with IoT is described. The section IV and V discusses IoT application areas and smart home application, respectively. The research directions, challenges in smart home, IoT data analytics and conclusion is given in section VI, VII, last section respectively.

Figure 1. Growth of IoT market (in billion dollars) by statista.com (Forecast end-user spending on IoT solutions worldwide from 2017 to 2025, n.d.) accessed on 30 December'2020

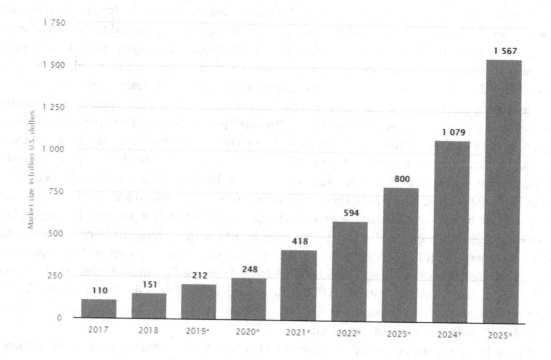

2 INTERNET OF THINGS

Internet of Things (IoT) aim is to give a smarter environment and simple life in less time, energy, and money. The IoT technologies can reduce overall running cost of many industries. The IoT have grown at fast speed in recent past and become a trend. IoT devices are connected to one another through internet and share data between each other to improve their performance (Siow et al., 2018). These actions occur automatically with minimum human interventions (European Commission, 2015). The International Telecommunication Union (ITU) calls the IoT "a global infrastructure for the information society, enabling advanced services by interconnecting things based on existing and evolving interoperable information and communication technologies" (International Telecommunication Union, 2012) and from a broader perspective, "a vision with technological and societal implications", which draws its language from the World Economic Forum report (World Economic Forum, 2012).

A typical IoT device can have four components: sensors, networks, data, and system monitoring. In IoT based systems connectivity is important and a necessary condition. Therefore, the internet protocols are key components of IoT technology and needs improvement. The main three components in communication protocols are:

(i)   *Device to device (D2D)* - In this method, communication take place between participating device such as mobile phones using cellular networks.

(ii) *Device to server(D2S)* – The collected heterogenous data reaches servers periodically from different participating devices. This scenario is based on cloud processing.

(iii) *Server to server(S2S)*– In this method, communication between servers takes place, servers exchange data between each other. This is possible only in cellular networks.

3 MACHINE LEARNING AND IOT

Machine learning along with IoT with deeper insights will create accurate, efficient, and cost-effective IoT devices without any limitations. The IoT analytics and machine learning gives real-time control of connected devices helping in fast response to potential issue. As mobile network shifting towards 5G technologies with higher throughput, it is expected that number of interconnected devices will increase in near future. The machine learning and Artificial Intelligence (AI) tools can gain insights from large scale IoT data and timely delivery of solutions to real world problems by reducing latency. IoT data analytics is gaining popularity to due to following reasons:

Massive Data Movement due to IoT Based Device

As per the report by Ericsson (Internet of things forecast mobility report, 2020), by 2050 more than 24 billion devices will be interconnected globally that includes everyday objects such as streetlights, gym vest, cars, water pumps, fitness trackers, electric meters and many more. The IoT based devices and their wide applications area has increased over time. Henceforth, the analysis of IoT data will play a major role in identifying the future needs by mining such heterogenous IoT data for prediction purpose.

Heterogeneous Sources and Variety of Data Types

There are large variety of IoT devices in demand such as laptops, tablets, mobile phones, short range devices and smart home devices. Such heterogeneity of data, the formats, structure, and attributes are making analytics even more difficult. For example, the IoT devices to be used for medical facilities and smart home environment can never be same. Therefore, storage, processing, and maintenance of this heterogenous data have become a challenging task. The paper (Adi et al., 2020) discusses similar issues related to heterogenous nature of data sources and giving insights over the critical questions.

IoT Stream and Uncertainty

The IoT data and analysis might face uncertainty issues due to failure of any IoT device or transmission during data transfer (Ed-daoudy & Maalmi, 2018). Such failure result in loss of data or error in data streams so compensate it advanced analytics or preprocessing techniques will be needed. Sometimes cyber intrusion is one of the reasons for such failure in transmission (Stellios et al., 2018). The model accuracy during data analytics will depend upon correct representation of uncertainties and propagation.

Scalability and Balancing

The collected IoT data is stored in cloud, various data analytics steps take place in the cloud. The transfer of IoT data to cloud platform is expensive and challenging too due to large number of IoT devices.

3.1 Data Analytics Method and Categories for IoT

In this section different data analytics techniques are discussed. Figure 2 shows the major components of analytics process.

Figure 2. Categories of data analytics on IoT data

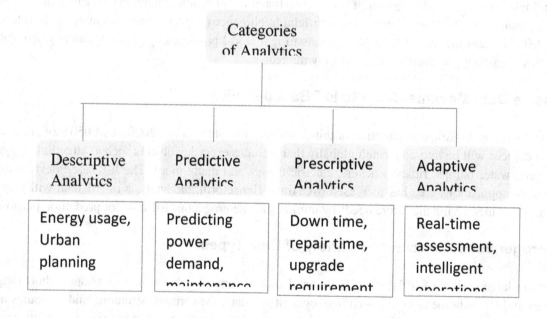

3.1.1 Descriptive Analytics

One can get insights into past events using advanced machine learning algorithms on data available on cloud environment of IoT based devices. The descriptive analytics is performed to monitor IoT devices status, applications, and services. To perform these analytics large volume of data is required, and that data can be stored in cloud environments. Many questions related to product usage, anomalies, throughput, and utilization can be answered by analysis. The classification techniques can be used to classify IoT data at different levels (Symeonides et al., 2019).

3.1.2 Predictive Analytics

Predictive analytics is performed on historical data using sophisticated mathematical or machine learning methods and model activities or patterns to forecast the potential trends or upcoming events. To sum up, it forecasts future events by learning the past dynamics and data associations of the current data (Predictive analytics history & current advances, n.d.). Different IoT sensors produce lots of data through real-time events such data is often considered too complex to handle. Such occurrences must be interpreted with minimum delay to be later used in decision-making in resolving the current situation (Akbar et al., 2017). The IoT systems, such as the traffic control system or the logistics in supply chain management of super stores, involves massive data that need to be processed in almost real-time to make decisions.

3.1.3 Prescriptive Analytics

Prescriptive analysis advises how to respond to potential incidents based on data analysis. The predictive analytics gives predictions, which further transformed into a recommended action to improve the situation. For instance, prescriptive analytics recommends shutting down the internet connection in case of unusual or suspicious traffic. It is considered a step towards optimizing decision making by suggesting required changes saving lots off efforts. In IoT technologies real-time streaming demands prescriptive analytics to give prescriptions for reliable and effective decision making (Lepenioti et al., 2020). The process incorporates predictive analytics predictions and utilizes artificial intelligence in a probabilistic context to provide automated, constrained, time-dependent and optimal decisions (Basu, 2013).

3.1.4 Adaptive Analytics

In the implementation process of predictive analytics, the results obtained needs adjustments with real data. The adaptive analytics optimizes and adjusts the process outcome by using recent history and underlying correlation in the real-time data. Adaptive analysis improves performance and reduces the errors when new input is received. The real-time assessment of data streams found in malware (Vu et al., 2019) can be performed using adaptive analytics.

3.2 IoT Data Analytics and Architecture

An important aspects IoT data processing is its architecture, and the popular ones are cloud and fog computing. IoT applications use both the frameworks depending on the application and process location. Different components of IoT and how data analytics is related to them is illustrated in Figure 3. The

Cloud server setup gives good results where high computing power is required but edge computing best fits in lightweight and online algorithms.

3.2.1 Cloud Computing

Cloud based environment utilizes internet to store, manage, and process large amount of data remotely. Cloud computing is decentralized paradigm that provides services that includes software, servers, databases and infrastructure. Different services such as Software as a Service (SaaS), Infrastructure as a Service (IaaS) and Platform as a Service (PaaS) are possible because of cloud computing technologies. Cloud computing is useful in IoT based environment, where massive data needs storage and analysis (Sharma & Wang, 2017). The Internet of Things and cloud platform fusion ensures that it will provide great access, connection and sharing of information between devices in the most effective way.

3.2.2 Edge Computing

In edge computing data processing takes place closer to devices or edge where data is produced. The processing of real-time data near device eliminating delay and saving the bandwidth (Tahsien et al., 2020). The computing process runs applications as close as possible to device generating data instead of cloud servers. For instance, when a car automatically calculates fuel consumption using data received by sensors in the car, this is known as edge computing. This process improves performance because large quantity of data does not need to be transferred to cloud for computation, computation is performed locally. The cloud computing is expensive because of process like connectivity, bandwidth, data migration and latency involved, so edge computing has an advantage over cloud computing. Edge computing analyzes data locally, therefore protects sensitive data and improve the speed of processing.

3.3.3 Fog Computing

The paper (Jain & Singhal, 2016) proposed the fog computing idea that also talks about the problem of latency in certain applications. Fog computing provides networking services between cloud and end-devices. In addition, it is suitable for applications with response time in range milliseconds to minutes, but fog nodes have limited data storage capacity (Fog computing and the internet of things: extend the cloud to where the things are, 2015). The fog is like a cloud closer to ground, which made possible new applications and services. For instance, the fog computing can deliver high quality streaming even in moving vehicles through access points positioned along highways (Bonomi et al., 2012). Various fog-based applications involve real-time communications rather than batch processing.

4 APPLICATIONS OF IOT DATA ANALYTICS

IoT based device produces massive amount of data that are valuable only if we perform analysis to gain insights in real-time. The data analytics (DA) is a process to discover useful insights and pattern from IoT data. In this section we discuss DA and its practicality in some application area.

Figure 3. Different architecture types for IoT data processing

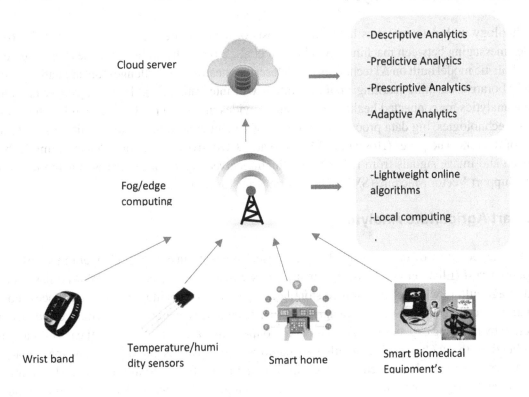

4.1 Smart Vehicles Analytics

IoT is changing the way vehicle can be used, data analytics can make decision making in vehicle environment efficient and accurate. The paradigm called, Internet of vehicle (IoV) (Contreras-Castillo et al., 2018) generates massive volume of messages to be interpreted. The businesses like to store such data for analytics to make improvements in service and product optimization. The IoV systems faces security issues in form of cyber anomalies or cyber intrusion attempts, intelligent solutions to combat such cyber-attacks is essential for providing seamless operations (Lee & Kim, 2018). A blockchain-based approach is discussed for ensuring security for stored data and for data distribution. The method defines that various nodes in vehicle environment could be integrated to form blockchain using blockchain architecture. Most of smart vehicles have high speed and rely on open wireless channel for communication causing disconnection leading to data faults (Zhang et al., 2018). The model takes spatial and temporal features into considerations for connected vehicle IoT data and proposes analytical framework that detects faults accurately (Zhang et al., 2018). Another framework for analytics is proposed in paper (He et al., 2014) that describes cloud based intelligent parking and vehicular data mining service.

A public transportation framework is proposed in paper (Luo et al., 2019), the author proposed solutions that can schedule subway, shared taxi, and bus effectively. The algorithm for pattern mining determines passenger and traffic flows periodically. The research (Kong et al., 2018) proposes an intelligent parking space allocation using auction-based mechanism.

4.2 Smart Healthcare Analytics

IoT technology with data analytics helps in diagnosis of health issues (Sakr & Elgammal, 2016) and provides messaging between machines (Salahuddin et al., 2018). The author proposes (Farahani et al., 2020) a holistic model built on AI technology and describes healthcare challenges that are patient centric. The collaborative machine learning implementation will integrate device layer, Fog layer and cloud. IoT data analytics for connected health care system provides increased resilience, seamless fusion with different technologies, big data processing, forecast of patient condition, health monitoring and accessibility of doctors. The paper (Hossain & Muhammad, 2018) gives an emotion detection model based on speech and image signals from IoT devices. Such received signals are processed using techniques such as Support Vector Machine (SVM) and Fourier transform to detect emotions.

4.3 Smart Agriculture Analytics

The data analytics over IoT data will enable smart agriculture that can deliver high operational efficiency and good harvest (Elijah et al., 2018). For many years wireless sensor network (WSMs) has been deployed in agriculture with a focus on precision agriculture, environmental monitoring, process control automation, and traceability. Analysis of agricultural data enhances operating performance and increases production by reducing expenditure. DA has been grouped into categories based on the specifications of IoT applications that include real-time, off-line, memory-level, business intelligence, and mass analytics. Image processing methods can detect disease in leaf, stem, and fruits, quality of fruits. The combination of IoT and image processing deployment produces high-quality produce and reduces crop failure. Different DA methods have been discussed in the paper (Marjani et al., 2017), the main methods are classification, prediction, clustering, and association rule.

The farmers can get help in different aspects of a farm such as irrigation system and storage management, also they can get information in advance about extreme weather condition or diseases that affects overall crop production. For example, to monitor fire outbreak in forests sensors may be planted at different locations, the sensors can also predict the region with the highest risk of fire. Some agricultural produce is damaged due to bad environmental factors in the warehouse. The factors such as temperature, moisture affect the quality of food and insects, rodents affect the quality of produce. The IoT and data analytics together will improve such storage facilities monitoring systems. An automatic decision system will monitor storage facilities and environmental conditions. The data collected can be analyzed to adjust environmental conditions and to send warning alerts to farmers when extreme conditions is reached. Data analytics measures collected data from sensors for use of the right quantity of chemicals and fertilizers in different regions of the farm. The right quantity of fertilizers will impact productivity as well as farming cost.

4.4 Smart Energy Systems Data Analytics

The energy companies analyze data for improving smart grids and for reducing recurring costs. The IoT technology has enabled innovative ways to build smart energy systems and deliver energy efficiency to customers by installing interconnected smart meters. Such smart meter records energy usage, voltage levels, and current for monitoring and billing purpose. The energy operations improve with proper data analytics and energy management becomes effective (Al-Ali et al., 2017). The author in (Alahakoon &

Yu, 2016) studied future possibilities of big data and cloud analytics for smart grids. The tools in the cloud platform perform advanced analytics on smart meter data, some popular tools are IBM Coremetrics (IBM digital analytics, 2020) and Google BigQuery (Bigquery: Cloud data warehouse, 2020). The author proposes (Siryani et al., 2017) a smart meter IoT system that predicts cost to be used by decision support systems in giving recommendations. By integrating machine learning with IoT based smart grid, the predictions related to power consumption, price, optimum schedule, fault detection, power generation, intrusion detection can be performed (Hossain et al., 2019). A theft detection system is proposed in paper (Li et al., 2017) that predicts the power consumption using Multilayer perceptron and recurrent neural network model. The usage history data decide if use is a theft or not.

4.5 Smart City Analytics

With increase in population and industrialization cities needs methods to handle it effectively. The main purpose of the development of the smart city is to improve traffic, water, energy management, and improvement in the quality of life of common citizens. In addition, smart cities will decrease expenses in health, transportation, safety, and resource management. The author claims that in long run a smart city various departments, government offices, various systems will monitor their own conditions and when required will self-repair themselves (Petrolo et al., 2017). A smart city will provide smart mobility through smart traffic control systems, autonomous cars, and public transportation (Von Hippel, 2005). A smart city influences many areas of human life such as environmental monitoring, social health, and crime monitoring. The decision related to city infrastructure, design, and development will be a lot easier with IoT technology. Through IoT and data analytics the authorities can take better decisions and predict which part of the city will be crowded in the future. The IoT technology and data analytics will enhance urban planning and infrastructure. Smart cities have several applications for AI-powered IoT-enabled technologies, from ensuring a cleaner atmosphere to developing public transport and safety. In the diagram below, you can see some of the smart city usage cases.

5 IoT and Secure Smart Home

A living space with dependencies on internet, the interconnection between devices such as heat, air conditioner, lightings, coffeemaker, security systems, and door locks that provide easy monitoring is called a smart home. Smart Home idea incorporates smart applications into everyday human life. Smart Homes is facing security and management problems due to the limited capability of small sensors, multiple Internet access applications and the heterogeneity of home networks that enable inexpert users to customize devices and microsystems.

5.1 Smart Home Components

The sensors, actuators, and mixed devices are three main components inside the smart home environment. The sensors, for example, light sensors, thermometers, or button switches feed real-world information into a smart home network. The light bulbs, coffee machines, and smart locks perform actions through the information over the internet. The mixed devices are powerful than actuators and sensors e.g., entertainment systems, smart speakers, or surveillance systems. In addition, home and other smart devices communicate with each other through a central gateway. Personal computers and smartphones are not considered smart home devices.

5.2 Smart Home Architecture

Smart home setup can be done in different ways, we discuss four main architecture used for smart homes.

Wi-fi enabled architecture- Different IoT devices are controlled using smart-phone applications that are user friendly with easy buttons so that even a non-technical person can use with ease. The IoT devices from same manufacturer can be controlled using same application (illustrated in Figure 4).

Figure 4. Wi-fi enabled smart home devices

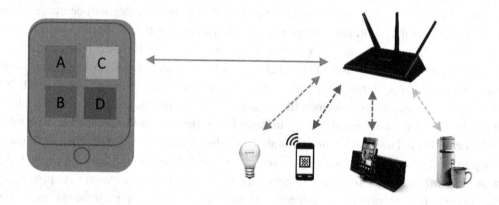

Integrated systems- Many modern homes have integrated systems built in the building that includes components to fulfill various needs of the house such as air-conditioning, heating, fire alarm, ventilation, and lighting. The control panel to interconnect such systems are physical devices located inside the house or building such a touch screen or sensors (illustrated in Figure 5). The control panel communicates with central server and requested actions is performed. The only problem is that it can rarely be extended or customized (Jacobsson et al., 2016).

Cloud based architecture- Some smart home devices are enabled through cloud-based support for which they need internet. The setup works through interface such as Alexa or Google Assistant and all dependent devices needs to be connected to these interfaces using user login. In most of cases only one application can control all the IoT devices inside smart home. The architecture of cloud-based smart home is illustrated in Figure 6.

Hub-based architecture- A hub is a central unit that connects all devices in the network to one single interface. In this type, devices communicate with each other without internet, but it requires a physical hub to be installed within the premises of a home or building. The advantage is safe data movement but increases the cost of smart home development.

Figure 5. Integrated Architecture for Smart home devices

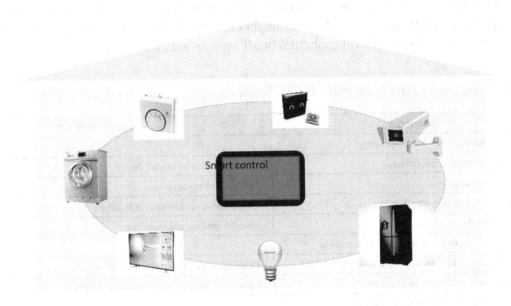

Figure 6. Cloud-based architecture for smart home

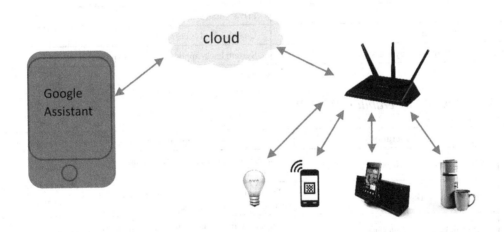

5.3 Security in Smart Home

The top ten vulnerabilities for IoT devices are compiled in (OWASP, n.d.) by Open Web Application Security Project (OWASP). Table 1 describes different types of risks and consequences to the smart home dwellers. The insecure control interfaces are a primary concern in both web-based architecture and smartphone applications. The attacker can hack into the system through the device, or communication channel, or through internet service providers. If the attacker gets access to the control panel, IoT devices control goes totally into his hands. They might even change the overall configuration to deny services to the owners. The top ten vulnerabilities in IoT systems are given in Table 2.

Table 1. Different types of risk in IoT systems (OWASP, n.d.)

Description	Consequence/impact
Threats caused through software components	
Improper authentication in the gateway	Unauthorized access in IoT device
Incomplete accountability	Unregistered system events
Inadequate software security measures in apps	Unauthorized modification of functions
Software vulnerability in interface	Unauthorized use and tempering
Threat originating through information or input process	
Access control and authentication insufficiency	Duplication, surveillance, manipulation and removal
Improper Access control policy and configuration	Incomplete authentication and access control
Threats coming through communication channel	
Authentication and confidentiality issues	Duplication, surveillance, deletion, data theft
Incomplete authentication and confidentiality in cloud	Duplication, surveillance, deletion, data theft, misuse
Threats caused by end users	
Untrained end-user, novice user etc.	Social engineering
Gullible users	Privacy issues and threats
Weak password choice by user	Unauthorized access
Poor systems configuration, setup by users	System hack, application hack, exploration attacks

Table 2. OSWAP Top 10 Vulnerabilities (2018)

Rank	Vulnerability	Examples	Impact
1	Weak, guessable passwords	unauthorized access, publicly available credentials	severe
2	Insecure network services	Unused ports, compromise confidentiality, availability of information	severe
3	Insecure Ecosystem interfaces	Lack of authentication and authorization, lack of input/output filtering	severe
4	Lack of secure update mechanism	Lack of anti-rollback, notification of security changes	moderate
5	Use of outdated components	Compromised supply chain	moderate
6	Insufficient privacy protection	Personal data not secure	severe
7	Insecure data transfer and storage	Lack of encryption, lack of access control	severe
8	Lack of device management	Poor service, lack of security support	Moderate
9	Insecure default setting	Difficulty in modifying configuration	moderate
10	Lack of physical hardening	Gain sensitive information, remote attack, local control	severe

6 CHALLENGES IN NEXT GENERATION SMART HOME

The development in automation and IoT made its way into home environment and security. Many considers that IoT provide peace of mind to the user, but security of smart home is still a challenging issue (Lin & Bergmann, 2016). In this section we discuss some existing challenges and methods to make a secure smart home environment.

Privacy in Smart Home

The smart home can be an easy target by cyber attackers and emerges as the major concern for system designers. In recent times different types of data are collected for data analytics purposes without ensuring who will receive and process data (Scott & Ketel, 2016). The devices such as Apple iWatch, Google Glass, Apple home kit, and Google fit gathers user data like financial and health details (Abomhara & Køien, 2014; Werth et al., 2020). Moreover, the user is constantly feeding information to apps running on devices without being fully aware of situations and consequences. For instance, whenever a user turns on a smart light using the app, he gives away the information about his presence at home or when he leaves his home, etc. To make a smart home environment and their user protected smart privacy mechanisms are required. Several studies show that the development of effective privacy mechanisms at secure smart home is a challenge due to large data collected by IoT devices (Perera et al., 2015).

Data Storage and Analytics

The data generated from IoT devices are massive, arrive at real time and variable in nature. The data analytics and storage of this massive data is complex but give insights about data that might be useful in improving services. The conventional method such as relational database management systems are not suitable for the task (Zaslavsky et al., 2013). The IoT data can be stored and managed by Cloud Technology and will provides advanced complex analysis (Misra et al., 2015).

Interoperability Issues

The development of a smart home system needs different IoT devices for specific needs of the owner. Use of such devices makes the solution costly and interoperability becomes one of the issues in open market. The companies in IoT market are producing diverse range of smart devices and want to achieve full interoperability to ensure smooth integration with internet. Z-wave products are interoperable and ZigBee Alliance with its Zigbee 3.0 has implemented interoperability. The Zigbee committees define properties for different vendors to develop interoperable devices for profiles like Home automation, health care, remote control etc (Fouladi & Ghanoun, 2013). The solutions are required to handle interoperability issues within smart home environment.

Multiple Devices and Apps in Smart Home System

In smart home systems, there are several sensors and small devices that are monitored using the application. To use a smart home device, users need an application to operate which makes the process complex and control complicated. The lack of communication between smart home devices is another weak point to be addressed. If a customer has two separate brands of smart devices for his house, for example, Alexa and Nest. Various applications need to be enabled to monitor the system and allow communication. This compounds the problems to the users, if he wants to control a device, he must launch a new app.

Hardware Limitation

The home devices and hardware limitation make the environment vulnerable against cybersecurity attacks. Another limitation is positioning of camera and motion sensors on accuracy. Security issues still exist in a smart home even if many algorithms and mechanisms are deployed. Manual adjustment in some devices is another concern that needs researcher attention.

Financial Cost for Smart Home

The IoT implementation cost for smart home depends on hardware, software and required services. The secure smart home system is costly due to use of many devices and technologies. To provide a low-cost implementation research is very much required so that it becomes affordable in terms of device cost and their annual maintenance.

7 TRENDS AND OPEN ISSUES IN IOT DATA ANALYTICS

Data analytics over IoT data contributes to improving services and applications. IoT must prevail over existing challenges to benefit from data analysis and gain insights. The main challenges in data analytics of IoT data are described in the following points.

Data Characteristics and Complexity

The IoT data is basis of analytics tools so the quality of the IoT data is extremely important. The IoT devices produce a high volume, fast velocity, and diverse range of data. Storing, managing, and preserving the data quality is a challenging task. A wide variety of solutions exists to extract, filter, and preprocess such data but these need improvement. For example, semantic methods enhance the abstraction of IoT data through annotation algorithms, but they require improvement so that velocity and volume is handled well. The data comes in all forms structured and unstructured that keeps on increasing making it more complex.

Diverse Applications

There are different categories of IoT applications based on different attributes and features. The issues from specific domain should be considered before running data analytics so that accurate results are obtained. The privacy of collected data is highly important. For example, if someone wants to use data from healthcare devices, he needs prior permission from stakeholder.

Algorithms for Analytics

As per the characteristics of IoT data, analytics algorithms are capable to process massive amount of data and extracts useful insights from them. The analysis requires advanced machine learning algorithms to process data from variety of sources in real-time. The machine learning algorithms gives accuracy in processing data and discovering information from data depending on time. Such algorithm is influenced by the impurity and incompleteness in data. But in some cases, a data scientist faces difficulty in effective interpretation of the results. Further, semi-supervised algorithms that are trained on small, labelled data with a large amount of unlabeled data assist data analytics.

Distribution and Interoperability

The variety of data in IoT applications is quite diverse and a less researched aspect. The data comes from diverse range of sources such as social media, image data, surveillance videos, fitness data, geospatial data, speech data and personal data. To process such variety and extraction of knowledge is main challenge before manufacturer of IoT devices. The corroboration of all kinds of data is required to perform predictive analytics and to produce correct prediction (Rakhmawati & Hausenblas, 2012; Rakhmawati et al., 2013).

Volume and Velocity of IoT Data

The surge in IoT data gives rise to performance and scalability issue as cloud is not suited enough to process it faster in real-time. It is estimated that by 2025 almost 30% data will come through real-time sources and 95% of it will be from IoT. The size of IoT data is still a major factor in storage and processing that also affects the overall infrastructure cost. One solution to this problem is distributed analytics that can efficiently handle large-scale IoT data, and there is lot of potential in this area.

8 CONCLUSION

The IoT technology is fast-moving forward and becoming an integral part of our lives. Machine learning will provide insights and value to the consumer of smart systems when integrated with IoT. The chapter further discusses the challenges due to machine-to-machine communications. Different aspects of IoT data analytics particularly challenges are the main criteria before performing IoT data analysis tasks. In addition, a secure smart home case was studied on various aspects such as architecture, security issues, and challenges. The IoT device manufacturer should consider the risk associated with usage data and should protect the user from data theft, data leaks, cyber-attack, and denial of service. The smart homeowner should understand the risk involved in using IoT applications and smart IoT devices. The challenges in IoT data analytics are related to data characteristics, algorithms, data volume, and interoperability. Unauthorized access, tempering, usage, and denial of service are some of the security risks associated with a smart home.

REFERENCES

Abomhara, M., & Køien, G. M. (2014, May). Security and privacy in the Internet of Things: Current status and open issues. In *2014 international conference on privacy and security in mobile systems (PRISMS)* (pp. 1-8). IEEE.

Adi, E., Anwar, A., Baig, Z., & Zeadally, S. (2020). Machine learning and data analytics for the IoT. *Neural Computing & Applications, 32*(20), 16205–16233. doi:10.100700521-020-04874-y

Akbar, A., Khan, A., Carrez, F., & Moessner, K. (2017). Predictive analytics for complex IoT data streams. *IEEE Internet of Things Journal, 4*(5), 1571–1582. doi:10.1109/JIOT.2017.2712672

Al-Ali, A. R., Zualkernan, I. A., Rashid, M., Gupta, R., & Alikarar, M. (2017). A smart home energy management system using IoT and big data analytics approach. *IEEE Transactions on Consumer Electronics, 63*(4), 426–434. doi:10.1109/TCE.2017.015014

Alahakoon, D., & Yu, X. (2016) Smart electricity meter data intelligence for future energy systems: a survey. *IEEE Trans IndInform, 12*(1), 425–436. .2414355 doi:10.1109/TII.2015

Basu, A. T. A. N. U. (2013). Five pillars of prescriptive analytics success. *Anal. Mag., 8*, 8–12.

Bigquery: Cloud data warehouse. (2020). https://cloud.google.com/bigquery-ml/docs/

Bonomi, F., Milito, R., Zhu, J., & Addepalli, S. (2012, August). Fog computing and its role in the internet of things. In *Proceedings of the first edition of the MCC workshop on Mobile cloud computing* (pp. 13-16). 10.1145/2342509.2342513

Contreras-Castillo, J., Zeadally, S., & Guerrero-Iban~ez, J. A. (2018). Internet of vehicles: Architecture, protocols, and security. *IEEE Internet of Things Journal, 5*(5), 3701–3709. doi:10.1109/JIOT.2017.2690902

Ed-daoudy, A., & Maalmi, K. (2018, November). Application of machine learning model on streaming health data event in real-time to predict health status using spark. In *2018 International Symposium on Advanced Electrical and Communication Technologies (ISAECT)* (pp. 1-4). IEEE. 10.1109/ISAECT.2018.8618860

Elijah, O., Rahman, T. A., Orikumhi, I., Leow, C. Y., & Hindia, M. N. (2018). An overview of Internet of Things (IoT) and data analytics in agriculture: Benefits and challenges. *IEEE Internet of Things Journal, 5*(5), 3758–3773. doi:10.1109/JIOT.2018.2844296

European Commission. (2015). *Digital Agenda for Europe: The Internet of Things.* https://goo.gl/oNhYOP

Farahani, B., Barzegari, M., Aliee, F. S., & Shaik, K. A. (2020). Towards collaborative intelligent IoT eHealth: From device to fog, and cloud. *Microprocessors and Microsystems, 72*, 102938. doi:10.1016/j.micpro.2019.102938

Fog computing and the internet of things: extend the cloud to where the things are. (2015). https://www.cisco.com/c/dam/en_us/solutions/trends/iot/docs/computing-overview.pdf

Forecast end-user spending on IoT solutions worldwide from 2017 to 2025. (n.d.). https://www.statista.com/statistics/976313/global-iot-market-size

Fouladi, B., & Ghanoun, S. (2013). Security evaluation of the Z-Wave wireless protocol. *Black hat USA, 24*, 1-2.

He, W., Yan, G., & Xu, L. D. (2014). Developing vehicular data cloud services in the iot environment. *IEEE Transactions on Industrial Informatics, 10*(2), 1587–1595. doi:10.1109/TII.2014.2299233

Hossain, E., Khan, I., Un-Noor, F., Sikander, S. S., & Sunny, M. S. H. (2019). Application of big data and machine learning in smart grid, and associated security concerns: A review. *IEEE Access: Practical Innovations, Open Solutions, 7*, 13960–13988. doi:10.1109/ACCESS.2019.2894819

Hossain, M. S., & Muhammad, G. (2018). Emotion-aware connected healthcare big data towards 5G. *IEEE Internet of Things Journal, 5*(4), 2399–2406. doi:10.1109/JIOT.2017.2772959

IBM digital analytics. (2020). https://www.ibm.com/in-en/analytics

International Telecommunication Union. (2012). *Overview of the Internet of Things*. Technical Report. International Telecommunication Union. https://www.itu.int/ITU-T/recommendations/rec.aspx?rec=11559

Internet of things forecast mobility report. (2020). https://www.ericsson.com/en/mobility-report/reports

Jacobsson, Boldt, & Carlsson. (2016). A risk analysis of a smart home automation system. *Future Generation Computer Systems, 56*(C), 719 – 733.

Jain, A., & Singhal, P. (2016, November). Fog computing: Driving force behind the emergence of edge computing. In *2016 International Conference System Modeling & Advancement in Research Trends (SMART)* (pp. 294-297). IEEE. 10.1109/SYSMART.2016.7894538

Karagiannis, V., Chatzimisios, P., Vazquez-Gallego, F., & Alonso-Zarate, J. (2015). A survey on application layer protocols for the internet of things. *Transaction on IoT and Cloud Computing, 3*(1), 11-17.

Kong, X. T., Xu, S. X., Cheng, M., & Huang, G. Q. (2018). IoT-enabled parking space sharing and allocation mechanisms. *IEEE Transactions on Automation Science and Engineering, 15*(4), 1654–1664. doi:10.1109/TASE.2017.2785241

Lee, C. H., & Kim, K. H. (2018, January). Implementation of IoT system using block chain with authentication and data protection. In *2018 International Conference on Information Networking (ICOIN)* (pp. 936-940). IEEE. 10.1109/ICOIN.2018.8343261

Lepenioti, K., Bousdekis, A., Apostolou, D., & Mentzas, G. (2020). Prescriptive analytics: Literature review and research challenges. *International Journal of Information Management, 50*, 57–70. doi:10.1016/j.ijinfomgt.2019.04.003

Li, W., Logenthiran, T., Phan, V. T., & Woo, W. L. (2017). Housing development building management system (hdbms) for optimized electricity bills. *Transactions on Environment and Electrical Engineering, 2*(2), 64–71. doi:10.22149/teee.v2i2.113

Lin, H., & Bergmann, N. W. (2016). IoT privacy and security challenges for smart home environments. *Information (Basel), 7*(3), 44. doi:10.3390/info7030044

Luo, X. G., Zhang, H. B., Zhang, Z. L., Yu, Y., & Li, K. (2019). A new framework of intelligent public transportation system based on the Internet of Things. *IEEE Access: Practical Innovations, Open Solutions, 7*, 55290–55304. doi:10.1109/ACCESS.2019.2913288

Manyika, J., Chui, M., Bisson, P., Woetzel, J., Dobbs, R., Bughin, J., & Aharon, D. (2015). *The Internet of Things: Mapping the value beyond the hype.* Academic Press.

Manyika, J., Chui, M., & Bughin, J. (2013). *Disruptive Technologies: Advances That Will Transform Life, Business, And the Global Economy.* Technical Report. McKinsey Global Institute. https://www.mckinsey.com/ business-functions/digital-mckinsey/our-insights/disruptive-technologies

Marjani, M., Nasaruddin, F., Gani, A., Karim, A., Hashem, I. A. T., Siddiqa, A., & Yaqoob, I. (2017). Big IoT data analytics: Architecture, opportunities, and open research challenges. *IEEE Access: Practical Innovations, Open Solutions, 5*, 5247–5261. doi:10.1109/ACCESS.2017.2689040

Misra, P., Rajaraman, V., Dhotrad, K., Warrior, J., & Simmhan, Y. (2015). *An Interoperable Realization of Smart Cities with Plug and Play based Device Management.* arXiv preprint arXiv:1503.00923.

OWASP. (n.d.). *Top IoT vulnerabilities - OWASP.* Available: https://www.owasp.org/index.php/Top_IoT_Vulnerabilities

Perera, C., Ranjan, R., Wang, L., Khan, S. U., & Zomaya, A. Y. (2015). Big Data Privacy in the Internet of Things Era. *IT Professional, 3*(17), 32–39. doi:10.1109/MITP.2015.34

Petrolo, R., Loscrì, V., & Mitton, N. (2017). towards a smart city based on cloud of things, a survey on the smart city vision and paradigms. *Transactions on Emerging Telecommunications Technologies, 28*(1), e2931. doi:10.1002/ett.2931

Predictive analytics history & current advances. (n.d.). *SAS.* https://www.sas.com/en_au/insights/analytics/predictive-analytics. html

Rakhmawati, N. A., & Hausenblas, M. (2012). On the Impact of Data Distribution in Federated SPARQL Queries. *Proceedings of 6th IEEE International Conference on Semantic Computing.* 10.1109/ICSC.2012.72

Rakhmawati, N. A., Umbrich, J., Karnstedt, M., Hasnain, A., & Hausenblas, M. (2013). *Querying Over Federated SPARQL Endpoints - A State of the Art Survey.* Technical Report. Digital Enterprise Research Institute. https://arxiv.org/abs/1306.1723

Sakr, S., & Elgammal, A. (2016). Towards a comprehensive data analytics framework for smart healthcare services. *Big Data Research, 4*, 44–58. doi:10.1016/j.bdr.2016.05.002

Salahuddin, M. A., Al-Fuqaha, A., Guizani, M., Shuaib, K., & Sallabi, F. (2018). *Softwarization of internet of things infrastructure for secure and smart healthcare.* arXiv preprint arXiv:1805.11011.

Scott, D., & Ketel, M. (2016, March). Internet of Things: A useful innovation or security nightmare? In *SoutheastCon 2016* (pp. 1–6). IEEE. doi:10.1109/SECON.2016.7506665

Sharma, S. K., & Wang, X. (2017). Live data analytics with collaborative edge and cloud processing in wireless iot networks. *IEEE Access: Practical Innovations, Open Solutions*, *5*, 4621–4635. doi:10.1109/ACCESS.2017.2682640

Siow, E., Tiropanis, T., & Hall, W. (2018). Analytics for the internet of things: A survey. *ACM Computing Surveys*, *51*(4), 1–36. doi:10.1145/3204947

Siryani, J., Tanju, B., & Eveleigh, T. J. (2017). A machine learning decision-support system improves the internet of things' smart meter operations. *IEEE Internet of Things Journal*, *4*(4), 1056–1066. doi:10.1109/JIOT.2017.2722358

Stellios, I., Kotzanikolaou, P., Psarakis, M., Alcaraz, C., & Lopez, J. (2018). A survey of iot-enabled cyberattacks: Assessing attack paths to critical infrastructures and services. *IEEE Communications Surveys and Tutorials*, *20*(4), 3453–3495. doi:10.1109/COMST.2018.2855563

Symeonides, M., Trihinas, D., Georgiou, Z., Pallis, G., & Dikaiakos, M. (2019, June). Query-driven descriptive analytics for IoT and edge computing. In *2019 IEEE International Conference on Cloud Engineering (IC2E)* (pp. 1-11). IEEE. 10.1109/IC2E.2019.00-12

Tahsien, S. M., Karimipour, H., & Spachos, P. (2020). Machine learning based solutions for security of Internet of Things (IoT): A survey. *Journal of Network and Computer Applications*, *161*, 102630. doi:10.1016/j.jnca.2020.102630

Von Hippel, E. (2005). Democratizing innovation: The evolving phenomenon of user innovation. *J. für Betriebswirtschaft*, *55*(1), 63–78. doi:10.100711301-004-0002-8

Vu, D. L., Nguyen, T. K., Nguyen, T. V., Nguyen, T. N., Massacci, F., & Phung, P. H. (2019). HIT4Mal: Hybrid image transformation for malware classification. *Transactions on Emerging Telecommunications Technologies*, 3789.

Werth, O., Guhr, N., & Breitner, M. H. (2020). Smart Home in Private Households: Status Quo, Discussion, and New Insights. *International Journal of Service Science, Management, Engineering, and Technology*, *11*(4), 122–136. doi:10.4018/IJSSMET.2020100108

World Economic Forum. (2012). *The Global Information Technology Report 2012 Living in a Hyperconnected World*. Technical Report. Author.

Zaslavsky, A., Perera, C., & Georgakopoulos, D. (2013). *Sensing as a service and big data*. arXiv preprint arXiv:1301.0159.

Zhang, H., Zhang, Q., Liu, J., & Guo, H. (2018). Fault detection and repairing for intelligent connected vehicles based on dynamic bayesian network model. *IEEE Internet of Things Journal*, *5*(4), 2431–2440. doi:10.1109/JIOT.2018.2844287

Zschörnig, T., Wehlitz, R., & Franczyk, B. (2020, September). IoT analytics architectures: challenges, solution proposals and future research directions. In *International Conference on Research Challenges in Information Science* (pp. 76-92). Springer. 10.1007/978-3-030-50316-1_5

Chapter 8
Smart Irrigation System for Crop Farmers in Namibia

Anton Limbo
University of Namibia, Namibia

Nalina Suresh
University of Namibia, Namibia

Set-Sakeus Ndakolute
University of Namibia, Namibia

Valerianus Hashiyana
University of Namibia, Namibia

Titus Haiduwa
University of Namibia, Namibia

Martin Mabeifam Ujakpa
International University of Management, Namibia

ABSTRACT

Farmers in Namibia currently operate their irrigation systems manually, and this seems to increase labor and regular attention, especially for large farms. With technological advancements, the use of automated irrigation could allow farmers to manage irrigation based on a certain crops' water requirements. This chapter looks at the design and development of a smart irrigation system using IoT. The conceptual design of the system contains monitoring stations placed across the field, equipped with soil moisture sensors and water pumps to maintain the adequate moisture level in the soil for the particular crop being farmed. The design is implemented using an Arduino microcontroller connected to a soil moisture sensor, a relay to control the water pump, as well as a GSM module to send data to a remote server. The remote server is used to represent data on the level of moisture in the soil to the farmers, based on the readings from the monitoring station.

DOI: 10.4018/978-1-7998-7541-3.ch008

INTRODUCTION

Agriculture plays an important role in the Namibian economy. A great number of the Namibian population depend on agricultural production for food and jobs. To ensure self-sufficiency and food security, the country needs to invest in the agricultural sector (Shikangalah & Mapani, 2020). The country is a sparsely covered environment with distinct seasons and limited areas that are good for pastoral and agricultural activities. Some areas are mountainous and arid, getting less than 50 millimetres of rainfall annually. Humidity is normally low, however rainfall is high in some areas in the northern and north eastern parts of Namibia (Liehr et al., 2016). Namibia being an arid country makes it very prone to the effects of climate change. These changes affect farming activities in the country, causing losses in the agriculture sector (Somses et al., 2020).

Most of the small scale farmers in the northern part of Namibia are communal farmers that grow crops such as pearl Millet, sorghum, and peanuts. These subsistence farmers survive on low rainfall and only produce what is enough for the household consumption and sell the surplus to make additional income (BDO, n.d.). Their crops are rain dependent, hence, become more vulnerable when drought occurs. However, some practice drip irrigation where water is directly poured at the crops and/or micro-irrigation where farmers ensure a low-volume sprinkler to the crops.

Commercial farmers in Namibia have a large area measured in hectares and at least own an irrigation system. The irrigation system allows them to control when and how to water crops in the field. However, this is done manually daily and based on human observations which is often not accurate (BDO, n.d.). Namibia has been listed by the World Resources Institute (WRI) as one of the top four countries that expect to have a significant increase in water stress by 2040 (Munyayi, 2015). Manual controlled irrigation systems or schemes currently in use in commercial farms are labor intensive, ineffective and waste a lot of already expensive and scarce water. Using these irrigation systems can result in either over-watering or under-watering. Namibia is arid to the semi-arid country with erratic rainfall and water scarcity has become a norm due to extreme weather. All four perennial rivers are found far at the northern and southern borders of the country (Liehr et al., 2018). It is a cumbersome process to derive water from such rivers and local dams due to varsity geographical location of crop farms. Urbanization led to a higher demand for water and not all farmers are well trained on water resources management. Indeed there are few innovative approaches towards water security in the country. For these reasons, there is an urgent need to find solutions to water crises to save businesses, farms and communities depended on the agricultural sector.

Modern technological advancements such as Internet of Things (IoT) can make it possible to smartly monitor and autonomously control the irrigation systems when farming. IoT refers to the interconnection of physical devices such as sensors and actuators using the internet (Wu et al., 2019). This interconnection enables IoT devices to be able to share information with each other or with users while enabling these users to monitor environments remotely. Adding Artificial Intelligence (AI) can enhance the use of IoT technologies. AI is the simulation of the human brain into machines, by making these machine act or perform tasks that human can perform (Lu et al., 2018). Today AI interacts with us in one way or another on a daily basis, starting with search engines such as Google to virtual assistants such as Siri and Alexa. AI has also been applied in many sectors of agriculture such as in disease control in animals and plants, pest control and soil treatment (Das et al., 2018). Special applications of AI have also been applied where a phenomena is understood by an AI system on a farm and having a decision taken based on the data about what is already known about this phenomena. This enables more optimal methods of managing farms as the more data is collected about the a phenomena, the more accurate the predictions become (Smith, 2020).

This chapter looks at applying IoT to automatically manage irrigation of crops in a field. The smart irrigation system is able to monitor and automate the irrigation process in a field. This is done by having monitoring stations across the field, equipped with soil moisture sensor, as well as the ability to autonomously start and stop the irrigation around the station using a water pump. The pump is activated and deactivated based on the readings from the soil moisture sensor. The monitoring stations are connected to a remote server which enables the farmers to observe the automation of irrigating crops while at the same time showing the status of the water pump at the stations, such as when the water pump was activated and for how long. This data can be used for future estimations of the water requirements of growing a particular crop. This chapter focuses on the development of a monitoring station prototype, while giving the full picture of how multiple monitoring stations can be deployed in a field.

The chapter concludes by proposing a fully automated smart agriculture architecture capable of monitoring other aspects in the field such as pests. The proposed architecture is capable of using AI to perform image recognition on the pests detected in the field and activate the spraying of a pesticide. Overall the developed prototype and the proposed smart architecture gives an idea of how future farming can be performed using modern technological advancements.

RELATED WORK

This approach has been done by other researchers, this section looks at previous studies done in the field which support the need for innovative ways to manage irrigation of crops.

Work done by Veena et al., (2013), implemented a real-time GSM-based automated irrigation control system using drip irrigation methodology. In this project, data was collected and processed using an ARM7 microcontroller. To provide a user interface, an android based application was developed to display data at the front- end and with that information, the user can decide what action to take. According to the command given by the user, solenoid valves control the irrigation. Their system supports aggressive water management for agricultural land. The architecture is based on the capabilities of current and next-generation microcontrollers and their application requirements. The microcontroller used for the system is promising that it can increase system life by reducing the power consumption. The automated irrigation system aimed to solve the same issue that is to automate irrigation without human intervention. However, this system does not have much security feature as it doesn't prompt users to provide any credentials to view collected data. The proposed solution addressed this drawback by developing a more secure mobile application that ensures user authentication.

Gutiérrez et al., (2014), developed an automated irrigation system using a wireless sensor network and GPRS module. The developed system aimed to optimize the water use for crops using two input parameters, namely: moisture sensor based on the electromagnetic measurement; and soil temperature sensor. The recorded data from sensors is stored locally in a memory chip and shared on a web-based application. A remote interface was provided through a GPRS module. The system worked on solar energy and tested in a sage crop field for 136 days. Results showed a saving of up to 90% of water compared to traditional systems. The limitation of this system is that when a user does not have internet access or smartphone to view the soil moisture and temperature level over the internet, there is no other way to get the soil moisture and temperature level. The IoT automated irrigation system had an integrated SMS platform that notifies the farmer regardless of the type of phone they are using. However, good network coverage is a strong requirement.

Ramachandran et al., (2018), demonstrated an Automated Irrigation System for Smart Agriculture using the IoT and cloud-connectivity to aggregate and store agricultural information. The experiments and simulations were used. The results showed that their optimization models can assist in reducing the water consumption during watering process.

A smart irrigation system was developed by Mérida García et al., (2018). The system employed solar energy to pump water for irrigating a olive orchard plot. This was done by developing a real-time Smart Photovoltaic Irrigation Manager (SPIM) which synchronized the photovoltaic power available with the energy required to irrigate the plot. The SPIM was equipped with different modules to calculate management aspects such as daily irrigation requirements as well as daily soil water ratio. Results showed that the SPIM was able to sufficiently provide water required to irrigate a crops in the field.

Work done by Goap et al., (2018) highlighted a smart architecture based on IoT capable of using weather forecasting as well as a machine learning algorithm to predict the moisture in the soil. The system used the rainfall forecasting and the predicted soil moisture to activate the irrigation system. The machine learning algorithm showed improved accuracy in predicting soil moisture, therefore leading to the efficient use of water to irrigate crops.

Zhao et al., (2017) implemented a Wide Area Network (WAN) based irrigation system using low-power, wide are network protocol (LoRa). The system contains an irrigation node capable of sending data to the cloud using LoRa gateways. LoRa protocol enabled irrigation nodes to communicate with the gateway device for a range of 8km. This gives this automation system the ability to cover a large field with a few irrigation nodes.

Hashiyana et al. (2020) implemented a co-design of an agricultural management android-based mobile application for small-scale farmers that would maximize farmer's crop production. This application assisted farmers to access agricultural information at hand without intervening agricultural officers. In addition, the prevalence, affordability and extensive use of smart phones has led to managing farming operations at ease and enhancement of agribusiness.

According Faustine et al. (2014) water shortages on farms are mainly caused by uncontrolled ways of monitoring water, therefore constant monitoring of water levels in the tank is highly recommend. Smart Water Level Monitoring System (SWLMS) was designed that will monitor water levels on a regular basis and provides a smart way to manage water resources on farms in the most cost-effective and convenient manner for farmers (Suresh et al., 2019). In addition, farmers will be able to monitor the water levels from any location at any given time and the system was able to monitor the tank water status and send out SMS notification when water level was at critical level and holistic view when the water level is at 40cm of 200cm which is 20% (Suresh et al., 2019).

SYSTEM DESIGN

The conceptual design of this automated irrigation system shows monitoring stations, spread out around the field. Each monitoring station contains a sensor to monitor the moisture in the soil. A pump is also placed at each monitoring station, this ensure that the soil moisture to be maintained is independent to that of other monitoring stations. This is done because they might be varying environmental aspects at locations of the monitoring stations even though they are in the same field. For example, Monitoring Station-1 might be in the sunlight while Station-2 is in the shade, meaning that there will be more evaporation at Station-1 compared to Station-2. The stations employ Arduino microcontroller as the main computing device at the stations.

Figure 1. System conceptual design

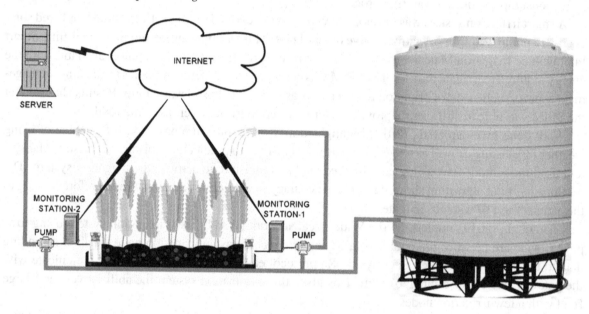

The sensor are then connected to the Arduino which also has a Global System for Mobile Communication (GSM) module connected to it, used to transmit the soil moisture readings to the Arduino microcontroller and analyzes the data comparing it to the set threshold. The data on the current level of moisture in the soil and status of the pump is sent to a remote server. The server then formats these readings from the multiple monitoring stations to give the farmer a representation of the moisture levels in the field as well as timeline of the levels of this moisture, while at the same time showing how long the pump was activated. Farmers access this data by logging into web server to view information on the moisture levels in the field.

Monitoring stations can also be configured to maintain a different level of moisture in the soil. This is to ensure that the irrigation system is adaptable based on the irrigation requirements of the crop being farmed as well as the availability of irrigation water. This ensures that the crop being grown is not under or over irrigated, which might result in the inadequate growth.

The soil moisture was measured in percentages, 1% for complete dryness of soil, 100% for high water content in the soil. When the soil moisture was less than or equal to the low threshold which was 20%, the relay allowed the water pump to receive power from the power source and irrigation starts. When the soil moisture was greater than or equal to the high threshold which was 60%, the relay stopped the water pump from receiving power from the power source and irrigation stopped. This ensures that the crop being grown is not under or over irrigated, which might result in the inadequate growth.

SYSTEM IMPLEMENTATION

To turn the conceptual design into a functional system, a protype was implemented as a monitoring station. The prototype employs low cost, low power consumption hardware such as an Arduino UNO microcontroller as main computing device for the monitoring station. A soil moisture sensor is attached to the Arduino to get periodic moisture readings from the soil. A GSM module serves as the communication device enabling the Arduino to send data to the server. A water pump is used to irrigate the soil by being controlled by the Arduino using an electoral Relay. The Relay manages the pump by opening power supply when the soil moisture reading is below 50% and cutting the power supply to pump when the reading reaches 80%. The actuator then activates water pump motor which pump water from the water container.

Arduino Integrated Development Environment was used to program the Arduino microcontroller. Other software used included Visual Studio Code, Sublime Text Code Editor and Node.js. For analytical purposes and timely information, a website was developed which formed as an interface between the system and the user (farmers). This platform allows the user to view the soil moisture condition in real-time. Data gathered from sensors is automatically sorted and converted to excel sheet and be encoded by Node.js and displayed on the web page's dashboard in the form of charts. Data visualization in the form of infographics helps farmers easily understand the interpreted data.

Figure 2. Implemented prototype

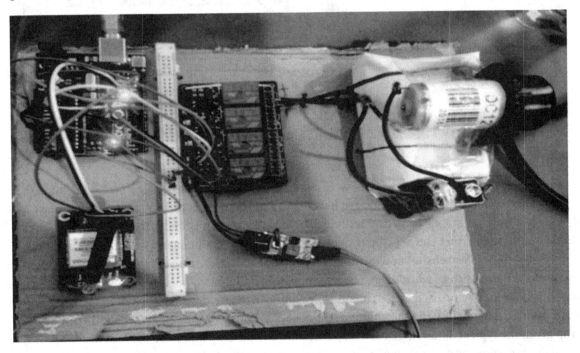

RESULTS AND DISCUSSION

The developed system prototype consisted of a monitoring station feeding information to a web server, which enabled the users to be able to monitor soil moisture in real-time while observing how these readings fluctuate based on the other environmental aspects such as humidity, wind, and sunlight that affect the moisture of soil.

Figure 3. Website login screen

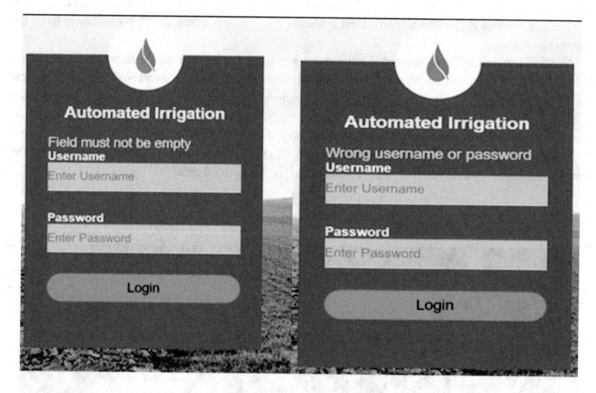

    Figure 3 shows a log in screen of the website. The idea here is that users (farmers) are able to view readings timely and also be able to observe how the soil moisture affects the growth of their crops. With this kind of observation, farmers can easily realize whether a crop is under irrigated or over irrigated. Based on this, the farmer can adjust the level of moisture to be maintained by the monitoring station.

    Figure 4 shows the website showing timely readings. The website also allows the farmers to view the status of the pump such as when the pump was activated and for how long. This can be used to calculate how much water is being used at each monitoring station in the field, which can be compared against the water source, therefore giving the ability to estimate the water consumption needed to grow a specific crop given the typical duration that the crop takes to fully develop and to be harvested.

    In the experimental setup shown in figure 5, a pump is connected to a water source, with the moisture sensor placed in a container with soil. This is to show how the monitoring station can be placed in the field as well as how to connect to the water source. The prototype monitors the soil moisture at regular intervals, it also provides a platform for farmers to monitor the soil moisture remotely.

Figure 4. Website showing timely moisture readings

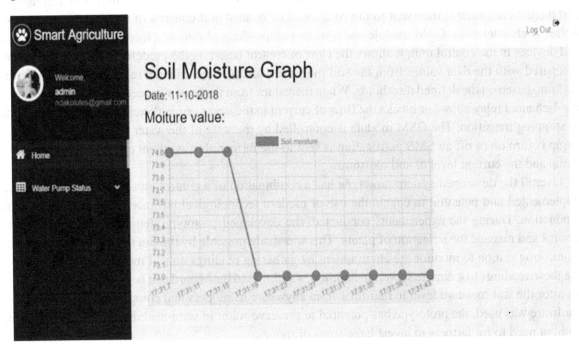

Figure 5. Experimental setup of the system

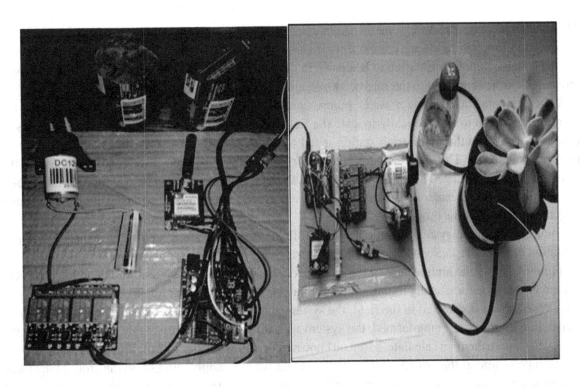

The sensing unit comprises of the soil moisture sensor which senses the content of water in the soil and the data acquired is then sent to the Arduino. The control unit consists of the Arduino microcontroller, 4-channel relay, GSM module and a water pump. The Arduino interfaces with the sensing unit and devices in the control unit, it allows the flow of control based on the predefined thresholds that are compared with the data values from the soil moisture sensor. The 4-channel relay is controlled by the Arduino based on the defined thresholds. When the values from the sensor exceed the defined threshold, the 4-channel relay allows or blocks the flow of current to the water pump therefore initiating irrigation or stopping irrigation. The GSM module is controlled by the state of the water pump. When the water pump is turn on or off an SMS notification is sent to the farmer alerting them of the state of the water pump and the current level of soil moisture.

Overall the developed system prototype has a combination of a control unit and a sensing unit was implemented and potential to enable the use of modern technological advancements to enhance crop production. During the experiments conducted, the developed prototype demonstrated the ability to control and manage the irrigation of plants. This was made possible by the use of IoT which enable the monitoring station to monitor the environment by gathering readings on soil moisture as well as sending these readings to a remote server. In addition, a web-based dashboard can be accessed by farmers to monitor the soil moisture level in real-time from anywhere using personal computer. Because low cost hardware was used, the prototype has potential to preserve water in semi-arid regions such as Namibia without need to for farmers to invest large sums of money.

CONCLUSION AND FUTURE WORK

This chapter looked at the development of a smart irrigation system for crop farming. The smart irrigation system used low cost hardware components such as an Arduino microcontroller as the main computing device on a monitoring station to be placed in the field. The monitoring station also included a soil moisture sensor, connected to the microcontroller, to read the level of moisture in the soil.

Other components include a water pump, controlled by the microcontroller which enables the monitoring station to automatically irrigate the field in the proximity area around the monitoring station. The data from the monitoring station is fed to a remote server using a GSM module connected to a microcontroller. This data includes readings from the soil moisture sensor as well as the time the water pump was opened. This data is represented to the user (farmer) using a webserver, and enables the farmer to estimate water consumption of the field as well as monitor how the level of soil moisture affects the growth of the crop.

Future work of the system include having a fully deployable system capable of managing not just irrigation but other tasks such as pesticides and fertilizers.

Figure 6 shows the architecture of the envisioned smart agriculture system. In the architecture, the system is able to automatically monitor the field for pests using Artificial Intelligence's image recognition. When a pest is detected in the field, the system automatically irrigates the pesticides in the field. Depending on the crop being farmed, the system will allow the farmer to simply enter the crop in the field. The system then can calculate, based on known information on the crop, the amount of water needed to adequately support the growth of the crop as well as when to apply fertilizer to the crop. Many more sensors can be added to the system such as pH sensors to enable the system to maintain an adequate pH balance needed for the adequate growth of the crop being farmed. The system can be integrated with other systems such as weather forecasting and instruments to manage the field. For example, irrigation can be synchronized with rainfall, to stop irrigating or to increase irrigation if the rainfall is not adequate for the required moisture level. Adding other instruments like Anemometers can be used to predict how the wind will affect the soil moisture.

Figure 6. Smart agriculture architecture

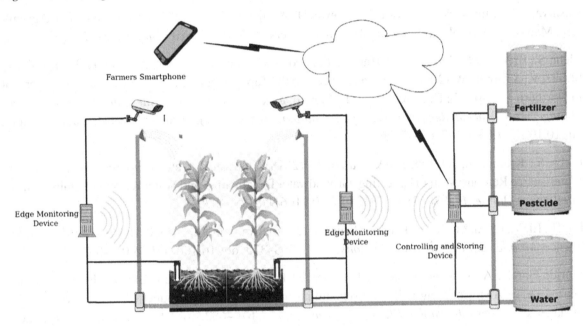

The envisioned system can enable farmers to grow crops in a much more automated approach. This will allow the farming of crops to be much more controlled, while at the same time maintaining the right conditions needed to adequately grow crops. The envisioned system has not been realized but it can be developed given the current technological advancements in fields such as Artificial Intelligence combined with Internet of Things devices.

REFERENCES

BDO. (n.d.). *Farming in Namibia*. Retrieved November 13, 2020, from https://www.bdo.com.na/en-gb/industries/natural-resources/farming-in-namibia

Das, S., Ghosh, I., Banerjee, G., & Sarkar, U. (2018). *Artificial Intelligence in Agriculture: A Literature Survey*. Academic Press.

Faustine, A., Mvuma, A. N., Mongi, H. J., Gabriel, M. C., Tenge, A. J., & Kucel, S. B. (2014). *Wireless Sensor Networks for Water Quality Monitoring and Control within Lake Victoria Basin: Prototype Development*. Wireless Sensor Network. doi:10.4236/wsn.2014.612027

Goap, A., Sharma, D., Shukla, A. K., & Rama Krishna, C. (2018). An IoT based smart irrigation management system using Machine learning and open source technologies. *Computers and Electronics in Agriculture, 155*, 41–49. doi:10.1016/j.compag.2018.09.040

Gutierrez, J., Villa-Medina, J. F., Nieto-Garibay, A., & Porta-Gandara, M. A. (2014). Automated Irrigation System Using a Wireless Sensor Network and GPRS Module. *IEEE Transactions on Instrumentation and Measurement, 63*(1), 166–176. doi:10.1109/TIM.2013.2276487

Hashiyana, V., Suresh, N., Haiduwa, T., Mbewe, D., & Ujakpa, M. M. (2020). Co-design of an Agricultural Management Application for Small-Scale Farmers. *2020 IST-Africa Conference (IST-Africa)*, 1–7.

Liehr, S., Brenda, M., Cornel, P., Deffner, J., Felmeden, J., Jokisch, A., Kluge, T., Müller, K., Röhrig, J., Stibitz, V., & Urban, W. (2016). From the Concept to the Tap—Integrated Water Resources Management in Northern Namibia. In D. Borchardt, J. J. Bogardi, & R. B. Ibisch (Eds.), *Integrated Water Resources Management: Concept, Research and Implementation* (pp. 683–717). Springer International Publishing. doi:10.1007/978-3-319-25071-7_26

Liehr, S., Kramm, J., Jokisch, A., & Müller, K. (2018). Integrated Water Resources Management in Water-scarce Regions: Water Harvesting, Groundwater Desalination and Water Reuse in Namibia. *Water Intelligence Online, 17*, 9781780407913. doi:10.2166/9781780407913

Lu, H., Li, Y., Chen, M., Kim, H., & Serikawa, S. (2018). Brain Intelligence: Go beyond Artificial Intelligence. *Mobile Networks and Applications, 23*(2), 368–375. doi:10.100711036-017-0932-8

Mérida García, A., Fernández García, I., Camacho Poyato, E., Montesinos Barrios, P., & Rodríguez Díaz, J. A. (2018). Coupling irrigation scheduling with solar energy production in a smart irrigation management system. *Journal of Cleaner Production, 175*, 670–682. doi:10.1016/j.jclepro.2017.12.093

Munyayi, B. S. (2015). Water Innovation: innovativee approaches towards water security in Namibia. Drfn.

Ramachandran, V., Ramalakshmi, R., & Srinivasan, S. (2018). An Automated Irrigation System for Smart Agriculture Using the Internet of Things. *2018 15th International Conference on Control, Automation, Robotics and Vision (ICARCV)*, 210–215. 10.1109/ICARCV.2018.8581221

Shikangalah, R. N., & Mapani, B. S. (2020). A review of bush encroachment in Namibia: From a problem to an opportunity? *Journal of Rangeland Science, 10*(3), 251–266.

Smith, M. J. (2020). Getting value from artificial intelligence in agriculture. *Animal Production Science, 60*(1), 46. doi:10.1071/AN18522

Somses, S., Bopape, M.-J. M., Ndarana, T., Fridlind, A., Matsui, T., Phaduli, E., Limbo, A., Maikhudumu, S., Maisha, R., & Rakate, E. (2020). Convection Parametrization and Multi-Nesting Dependence of a Heavy Rainfall Event over Namibia with Weather Research and Forecasting (WRF) Model. *Climate (Basel), 8*(10), 112. doi:10.3390/cli8100112

Suresh, N., Hashiyana, V., Kulula, V. P., & Thotappa, S. (2019). Smart Water Level Monitoring System for Farmers. In D. Goyal, S. Balamurugan, S.-L. Peng, & D. S. Jat (Eds.), *The IoT and the Next Revolutions Automating the World* (pp. 213–228). IGI Global. doi:10.4018/978-1-5225-9246-4.ch014

Veena, D., Ayush, A., Chandan, K., Raunak, R., & Rochak, B. (2013). A Real time implementation of a GSM based Automated Irrigation Control System using Drip Irrigation Metholgy. *International Journal of Scientific and Engineering Research*, *4*(5), 146–151.

Wu, D., Shi, H., Wang, H., Wang, R., & Fang, H. (2019). A Feature-Based Learning System for Internet of Things Applications. *IEEE Internet of Things Journal*, *6*(2), 1928–1937. doi:10.1109/JIOT.2018.2884485

Zhao, W., Lin, S., Han, J., Xu, R., & Hou, L. (2017). Design and Implementation of Smart Irrigation System Based on LoRa. *2017 IEEE Globecom Workshops (GC Wkshps)*, 1–6. doi:10.1109/GLO-COMW.2017.8269115

Compilation of References

Amruta, M. K., & Satish, M. T. (2013). Solar powered water quality monitoring system using wireless sensor network. *IEEE Conference on automation, computing, communication, control, and compressed sensing.* 10.1109/iMac4s.2013.6526423

Internet of things forecast mobility report. (2020). https://www.ericsson.com/en/mobility-report/reports

Raza, S., Misra, P., He, Z., & Voigt, T. (2015). Bluetooth smart: An enabling technology for the Internet of Things. *2015 IEEE 11th International Conference on Wireless and Mobile Computing, Networking and Communications (WiMob),* 155-162. 10.1109/WiMOB.2015.7347955

Wang, Z. L., & Wang, F. H. (2011). *Introduction to the internet of things engineering.* Mechanical Industry Press.

Zikopoulos, P., & Eaton, C. (2011). *Understanding big data: Analytics for enterprise class hadoop and streaming data.* McGraw-Hill Osborne Media.

Adi, E., Anwar, A., Baig, Z., & Zeadally, S. (2020). Machine learning and data analytics for the IoT. *Neural Computing & Applications, 32*(20), 16205–16233. doi:10.100700521-020-04874-y

Ertürk, M. A., Aydın, M. A., Büyükakkaşlar, M. T., & Evirgen, H. (2019). A Survey on LoRaWAN Architecture, Protocol and Technologies. *Future Internet, 11*(10), 216. doi:10.3390/fi11100216

Papageorgiou, P. (2003). *Literature Survey on wireless sensor networks. Report.* University of Maryland.

Rao, B. P., Saluia, P., Sharma, N., Mittal, A., & Sharma, S. V. (2012). Cloud computing for Internet of Things & sensing based applications. In *Sensing Technology (ICST), Sixth International Conference,* (pp. 374–380). IEEE. 10.1109/ICSensT.2012.6461705

Zhang, G. G., Bi, Y., & Li, C. (2013). Massive internet data security processing model research. *Small Microcomputer System, 34*(9), 2090–2094.

Ed-daoudy, A., & Maalmi, K. (2018, November). Application of machine learning model on streaming health data event in real-time to predict health status using spark. In *2018 International Symposium on Advanced Electrical and Communication Technologies (ISAECT)* (pp. 1-4). IEEE. 10.1109/ISAECT.2018.8618860

Fox, G. C., Kamburugamuve, S., & Hartman, R. D. (2012). Architecture and measured characteristics of a cloud based internet of things. In *Collaboration Technologies and Systems (CTS), International Conference,* (pp. 6–12). IEEE. 10.1109/CTS.2012.6261020

Koohi, I., & Groza, V. Z. (2014). Optimizing particle swarm optimization algorithm. *IEEE 27th Canadian Conference on Electrical and Computer Engineering.* 10.1109/CCECE.2014.6901057

Shafique, K., Khawaja, B., Sabir, F., Qazi, S., & Mustaqim, M. (2020). Internet of Things (IoT) For Next-Generation Smart Systems: A Review of Current Challenges. In *Future Trends and Prospects for Emerging 5G-IoT Scenarios*. IEEE Access. . doi:10.1109/ACCESS.2020.2970118

Yi, K. M. (2010). *Preliminary study of IoT security*. Internet Police Detachment of Public Security Bureau in Taiwan City.

Dash, S. K., Mohapatra, S., & Pattnaik, P. K. (2010). A Survey on Application of Wireless Sensor Network Using Cloud Computing. *International Journal of Computer science & Engineering Technologies*, *1*(4), 50–55.

Gupta, S. K., Singh, R. K., & Sharan, S. N. (2016). An approach to implement PSO to optimize outrage probability of coded cooperative communication with multiple relays. *Alexandria Engineering Journal*.

Lu, Y., Richter, P., & Lohan, E. S. (2018). *Opportunities and Challenges in the Industrial Internet of Things based on 5G Positioning*. . doi:10.1109/ICL-GNSS.2018.8440903

Stellios, I., Kotzanikolaou, P., Psarakis, M., Alcaraz, C., & Lopez, J. (2018). A survey of iot-enabled cyberattacks: Assessing attack paths to critical infrastructures and services. *IEEE Communications Surveys and Tutorials*, *20*(4), 3453–3495. doi:10.1109/COMST.2018.2855563

Zhang, L., & Wang, Z. (2006). Integration of RFID into wireless sensor networks: architectures, opportunities and challenging problems. In *Proceeding of the IEEE fifth international conference on grid and cooperative computing workshops GCCW '06* (pp. 463–469). 10.1109/GCCW.2006.58

Liu, W., Wang, Z., Liu, X., Zeng, N., & Bell, D. (2019). A Novel Particle Swarm Optimization Approach for Patient Clustering From Emergency Departments. *IEEE Transactions on Evolutionary Computation*, *23*(4), 632–644. doi:10.1109/TEVC.2018.2878536

Ning, Z., Xia, F., Ullah, N., Kong, X., & Hu, X. (2017). Vehicular social networks: Enabling smart mobility. *IEEE Communications Magazine*, *55*(5), 16–55. doi:10.1109/MCOM.2017.1600263

Sharma, H., Haque, A., & Jaffery, Z. (2019). *Smart Agriculture Monitoring using Energy Harvesting Internet of Things*. EH-IoT.

Symeonides, M., Trihinas, D., Georgiou, Z., Pallis, G., & Dikaiakos, M. (2019, June). Query-driven descriptive analytics for IoT and edge computing. In *2019 IEEE International Conference on Cloud Engineering (IC2E)* (pp. 1-11). IEEE. 10.1109/IC2E.2019.00-12

Zaslavsky, A., Perera, C., & Georgakopoulos, D. (2013). *Sensing as a service and big data*. arXiv preprint arXiv:1301.0159.

Hutson, M. (2017). A matter of trust. *Science*, *358*(6369), 1375–1377. doi:10.1126cience.358.6369.1375 PMID:29242328

Johan, J. (2018). Smart Soil Parameters Estimation System Using an Autonomous Wireless Sensor Network with Dynamic Power Management Strategy. *IEEE Sensors Journal, Volume*, *18*(21), 8913–8923. doi:10.1109/JSEN.2018.2867432

Predictive analytics history & current advances. (n.d.). *SAS*. https://www.sas.com/en_au/insights/analytics/predictive-analytics. html

Akbar, A., Khan, A., Carrez, F., & Moessner, K. (2017). Predictive analytics for complex IoT data streams. *IEEE Internet of Things Journal*, *4*(5), 1571–1582. doi:10.1109/JIOT.2017.2712672

Parwekar, P. (2011). From Internet of Things towards cloud of things. In *Computer and Communication Technology (ICCCT), 2nd International Conference on*, (pp. 329–333). IEEE. 10.1109/ICCCT.2011.6075156

Patil, K. A., & Kale, N. R. (2016). A model for smart agriculture using IoT. *IEEE International Conference on Global Trends in Signal Processing, Information Computing and Communication*, 543-545. 10.1109/ICGTSPICC.2016.7955360

Petrolo, R., Loscri, V., & Mitton, N. (2017). Towards a smart city based on cloud of things, a survey on the smart city vision and paradigms. *Transactions on Emerging Telecommunications Technologies*, *28*(1), e2931. doi:10.1002/ett.2931

Atzori, L., Iera, A., & Morabito, G. (2010). The Internet of Things: A survey. *Computer Networks*, *54*(15), 2787–2805. doi:10.1016/j.comnet.2010.05.010

Batty, M., Axhausen, K. W., Giannotti, F., Pozdnoukhov, A., Bazzani, A., Wachowicz, M., Ouzounis, G., & Portugali, Y. (2012). Smart cities of the future. *The European Physical Journal. Special Topics*, *214*(1), 481–518. doi:10.1140/epjst/e2012-01703-3

Lepenioti, K., Bousdekis, A., Apostolou, D., & Mentzas, G. (2020). Prescriptive analytics: Literature review and research challenges. *International Journal of Information Management*, *50*, 57–70. doi:10.1016/j.ijinfomgt.2019.04.003

Min, M., Wan, X., Xiao, L., Chen, Y., Xia, M., Wu, D., & Dai, H. (2019, June). Learning-Based Privacy-Aware Offloading for Healthcare IoT With Energy Harvesting. *IEEE Internet of Things Journal*, *6*(3), 4307–4316. doi:10.1109/JIOT.2018.2875926

Basu, A. T. A. N. U. (2013). Five pillars of prescriptive analytics success. *Anal. Mag.*, *8*, 8–12.

Elmaghraby, A. S., & Losavio, M. M. (2014). Cyber security challenges in smart cities: Safety, security and privacy. *Journal of Advanced Research*, *5*(4), 491–497. doi:10.1016/j.jare.2014.02.006 PMID:25685517

Ma, D., Lan, G., Hassan, M., Hu, W., & Das, S. K. (2020). Sensing, Computing, and Communications for Energy Harvesting IoTs: A Survey. IEEE Communications Surveys & Tutorials, 22(2), 1222-1250. doi:10.1109/COMST.2019.2962526

Vyas, Bhat, & Jha. (n.d.). IoT: Trends, Challenges and Future Scope. *International Journal of Computer Science & Communication*, *7*(1), 186-197.

Barreto, L., Amaral, A., & Pereira, T. (2017). Industry 4.0 implications in logistics: An overview. *Procedia Manufacturing*, *13*, 1245-1252. doi:10.1016/j.promfg.2017.09.045

Usman, M., Ahmed, I., Aslam, M. I., Khan, S., & Shah, U. A. (2017). *SIT: a lightweight encryption algorithm for secure internet of things.* arXivPrepr. arXiv 1704.08688

Vu, D. L., Nguyen, T. K., Nguyen, T. V., Nguyen, T. N., Massacci, F., & Phung, P. H. (2019). HIT4Mal: Hybrid image transformation for malware classification. *Transactions on Emerging Telecommunications Technologies*, 3789.

Al-Garadi, Mohamed, Al-Ali, Du, Ali, & Guizani. (2020). A Survey of Machine and Deep Learning Methods for Internet of Things (IoT) Security. *IEEE Communications Surveys & Tutorials*.

Ashton. (n.d.). That 'Internet of Thing' thing. *RFID Journal*.

Farooq, M., Waseem, M., Mazhar, S., Khairi, A., & Kamal, T. (2015). A Review on Internet of Things (IoT). *International Journal of Computers and Applications*, *113*, 1–7. doi:10.5120/19787-1571

Manyika, J., Chui, M., & Bughin, J. (2013). *Disruptive Technologies: Advances That Will Transform Life, Business, And the Global Economy.* Technical Report. McKinsey Global Institute. https://www.mckinsey.com/ business-functions/digital-mckinsey/our-insights/disruptive-technologies

Tao, F. (2014). CCIoT-CMfg: cloud computing and Internet of Things based cloud manufacturing service system. Academic Press.

Abdelhafidh, M., Fourati, M., Fourati, L. C., & Abidi, A. (2017). Remote Water Pipeline Monitoring System IoT-Based Architecture for New Industrial Era 4.0. *2017 IEEE/ACS 14th International Conference on Computer Systems and Applications (AICCSA)*, 1184-1191. 10.1109/AICCSA.2017.158

Ainane, N., Ouzzif, M., & Bouragba, K. (2018). Data security of smart cities. *ACM International Conference Proceeding Series.* 10.1145/ 3286606.3286866

Sharma, S. K., & Wang, X. (2017). Live data analytics with collaborative edge and cloud processing in wireless iot networks. *IEEE Access: Practical Innovations, Open Solutions, 5,* 4621–4635. doi:10.1109/ACCESS.2017.2682640

Suciu, G., Vulpe, A., Halunga, S., Fratu, O., Todoran, G., & Suciu, V. (2013). Smart Cities Built on Resilient Cloud Computing and Secure Internet of Things. In *Control Systems and Computer Science (CSCS), 19th International Conference on,* (pp. 513–518). IEEE. 10.1109/CSCS.2013.58

Abosaq, N. H. (2019). Impact of privacy issues on smart city services in a model smart city. *International Journal of Advanced Computer Science and Applications, 10*(2), 177–185. doi:10.14569/IJACSA.2019.0100224

Alagoz, F. (2010). From cloud computing to mobile Internet, from user focus to culture and hedonism: the crucible of mobile health care and wellness applications. In *ICPCA 2010.* IEEE. doi:10.1109/ICPCA.2010.5704072

Shirvanimoghaddam, M., Shirvanimoghaddam, K., Abolhasani, M. M., Farhangi, M., Zahiri Barsari, V., Liu, H., Dohler, M., & Naebe, M. (2019). Towards a Green and Self-Powered Internet of Things Using Piezoelectric Energy Harvesting. *IEEE Access: Practical Innovations, Open Solutions, 7,* 94533–94556. doi:10.1109/ACCESS.2019.2928523

Tahsien, S. M., Karimipour, H., & Spachos, P. (2020). Machine learning based solutions for security of Internet of Things (IoT): A survey. *Journal of Network and Computer Applications, 161,* 102630. doi:10.1016/j.jnca.2020.102630

Awad, A. I., Furnell, S., Hassan, A. M., & Tryfonas, T. (2019). Special issue on security of IoT – enabled infrastructures in smart cities. *Ad Hoc Networks, 92,* 101850. Advance online publication. doi:10.1016/j.adhoc.2019.02.007

Jain, A., & Singhal, P. (2016, November). Fog computing: Driving force behind the emergence of edge computing. In *2016 International Conference System Modeling & Advancement in Research Trends (SMART)* (pp. 294-297). IEEE. 10.1109/SYSMART.2016.7894538

Lohr, Sadeghi, & Winandy. (2010). Securing the e-health cloud. In *Proceedings of the 1st ACM International Health Informatics Symposium,* (pp. 220–229). ACM.

Yang, H., Kumara, S., Bukkapatnam, S., & Tsung, F. (2019). The Internet of Things for Smart Manufacturing: A Review. *IIE Transactions, 51*(11), 1–35. doi:10.1080/24725854.2018.1555383

Doukas, C., & Maglogiannis, I. (2012). Bringing iot and cloud computing towards pervasive healthcare. In *Innovative Mobile and Internet Services in Ubiquitous Computing (IMIS), Sixth International Conference on,* (pp. 922–926). IEEE. 10.1109/IMIS.2012.26

Fog computing and the internet of things: extend the cloud to where the things are. (2015). https://www.cisco.com/c/dam/en_us/solutions/trends/iot/docs/computing-overview.pdf

Meng, Y., Zang, W., Zhu, H., & Shen, X. S. (2018). Securing consumer IoT in the smart home: Architecture, challenges and countermeasures. *IEEE Wireless Communications, 25*(6), 53–59. doi:10.1109/MWC.2017.1800100

Tao, F., Zuo, Y., Xu, L. D., & Zhang, L. (2014, May). IoT-Based Intelligent Perception and Access of Manufacturing Resource Toward Cloud Manufacturing. *IEEE Transactions on Industrial Informatics, 10*(2), 1547–1557. doi:10.1109/TII.2014.2306397

Bonomi, F., Milito, R., Zhu, J., & Addepalli, S. (2012, August). Fog computing and its role in the internet of things. In *Proceedings of the first edition of the MCC workshop on Mobile cloud computing* (pp. 13-16). 10.1145/2342509.2342513

IoT Central. (n.d.). https://www.iotcentral.io/blog/iiot-protocols-for-the-beginners

Mitton, N., Papavassiliou, S., Puliafito, A., & Trivedi, K. S. (2012). Combining Cloud and sensors in a smart city environment. *EURASIP Journal on Wireless Communications and Networking*, *2012*(1), 1–10. doi:10.1186/1687-1499-2012-247

Siby, S., Maiti, R. R., & Tippenhauer, N. O. (2017). IoTscanner: Detecting privacy threats in IoT neighborhood. *Proceedings of the 3rd ACM International Workshop on IoT Privacy, Trust and Security*, 23-30.

Contreras-Castillo, J., Zeadally, S., & Guerrero-Iban~ez, J. A. (2018). Internet of vehicles: Architecture, protocols, and security. *IEEE Internet of Things Journal*, *5*(5), 3701–3709. doi:10.1109/JIOT.2017.2690902

Hassan, W. H. (2019). Current research on Internet of Things (IoT) Security: A Survey. *Computer Networks*, *148*, 283–294. doi:10.1016/j.comnet.2018.11.025

Kamilaris, A., Pitsillides, A., & Trifa, V. (2011). The smart home meets the web of things. *International Journal of Ad Hoc and Ubiquitous Computing*, *7*(3), 145. doi:10.1504/IJAHUC.2011.040115

Aazam, M., Huh, E. N., St-Hilaire, M., Lung, C. H., & Lambadaris, I. (2016). Cloud of Things: Integration of IoT with Cloud Computing. In A. Koubaa & E. Shakshuki (Eds.), *Robots and Sensor Clouds. Studies in Systems, Decision and Control* (Vol. 36). Springer. doi:10.1007/978-3-319-22168-7_4

Lee, C. H., & Kim, K. H. (2018, January). Implementation of IoT system using block chain with authentication and data protection. In *2018 International Conference on Information Networking (ICOIN)* (pp. 936-940). IEEE. 10.1109/ICOIN.2018.8343261

Leloglu, E. (2016). A review of security concern in Internet of Things. *J. Comput. Commun.*, *5*(01), 121–136. doi:10.4236/jcc.2017.51010

Ali, S., Bosche, A., & Ford, F. (2018). *Cybersecurity Is the Key to Unlocking Demand in the Internet of Things*. Bain and Company.

Wu, M., Tan, L., & Xiong, N. (2015). A Structure Fidelity Approach for Big Data Collection in Wireless Sensor Networks. *Sensors (Basel)*, *15*(1), 248–273. doi:10.3390150100248 PMID:25609045

Zhang, H., Zhang, Q., Liu, J., & Guo, H. (2018). Fault detection and repairing for intelligent connected vehicles based on dynamic bayesian network model. *IEEE Internet of Things Journal*, *5*(4), 2431–2440. doi:10.1109/JIOT.2018.2844287

Cui, Xie, Qu, Gao, & Yang. (2018). Security and Privacy in Smart Cities: Challenges and Opportunities. *IEEE Access*.

He, W., Yan, G., & Xu, L. D. (2014). Developing vehicular data cloud services in the iot environment. *IEEE Transactions on Industrial Informatics*, *10*(2), 1587–1595. doi:10.1109/TII.2014.2299233

Netlabtoolkit. (n.d.). https://www.netlabtoolkit.org/learning/tutorials/iot-cloud-services

Ismagilova, E., Hughes, L., Rana, N. P., & Dwivedi, Y. K. (2020). Security, Privacy and Risks within Smart Cities: Literature Review and Development of a Smart City Interaction Framework. *Information Systems Frontiers*, 1–22. doi:10.100710796-020-10044-1 PMID:32837262

Luo, X. G., Zhang, H. B., Zhang, Z. L., Yu, Y., & Li, K. (2019). A new framework of intelligent public transportation system based on the Internet of Things. *IEEE Access: Practical Innovations, Open Solutions*, *7*, 55290–55304. doi:10.1109/ACCESS.2019.2913288

Chen, Y., Zhao, S., & Zhai, Y. (2014). Construction of intelligent logistics system by RFID of Internet of things based on cloud computing. *Journal of Chemical and Pharmaceutical Research*, *6*(7), 1676–1679.

Haller, S., Karnouskos, S., & Schroth, C. (2009). The Internet of Things in an Enterprise Context. In J. Domingue, D. Fensel, & P. Traverso (Eds.), Lecture Notes in Computer Science: Vol. 5468. *Future Internet – FIS 2008. FIS 2008.* Springer. doi:10.1007/978-3-642-00985-3_2

La Rosa, R., Livreri, P., Trigona, C., Di Donato, L., & Sorbello, G. (2019). Strategies and Techniques for Powering Wireless Sensor Nodes through Energy Harvesting and Wireless Power Transfer. *Sensors (Basel)*, *19*(12), 2660. doi:10.339019122660 PMID:31212839

Li, J. (2019). A clustering based routing algorithm in IoT aware wireless mesh networks. *Sustainable Cities and Society*.

Manyika, J., Chui, M., Bisson, P., Woetzel, J., Dobbs, R., Bughin, J., & Aharon, D. (2015). *The Internet of Things: Mapping the value beyond the hype.* Academic Press.

Gubbi, J., Buyya, R., Marusic, S., & Palaniswami, M. (2013). Internet of Things (IoT): A vision, architectural elements, and future directions. *Future Generation Computer Systems*, *29*(7), 1645–166. doi:10.1016/j.future.2013.01.010

Kabir. (2020). *An overview of the Internet of Things (IoT) and IoT Security.* Research Gate.

Kong, X. T., Xu, S. X., Cheng, M., & Huang, G. Q. (2018). IoT-enabled parking space sharing and allocation mechanisms. *IEEE Transactions on Automation Science and Engineering*, *15*(4), 1654–1664. doi:10.1109/TASE.2017.2785241

Sakr, S., & Elgammal, A. (2016). Towards a comprehensive data analytics framework for smart healthcare services. *Big Data Research*, *4*, 44–58. doi:10.1016/j.bdr.2016.05.002

van Zoonen, L. (2016). Privacy concerns in smart cities. *Government Information Quarterly*, *33*(3), 472–480. doi:10.1016/j.giq.2016.06.004

Hossain, Fotouhi, & Hasan. (2015). Towards an Analysis of Security Issues, Challenges, and Open Problems in the Internet of Things. *IEEE World Congress on Services*.

Salahuddin, M. A., Al-Fuqaha, A., Guizani, M., Shuaib, K., & Sallabi, F. (2018). *Softwarization of internet of things infrastructure for secure and smart healthcare.* arXiv preprint arXiv:1805.11011.

Ali, M. A. (2019, December). IoT Security Evolution: Challenges and Countermeasures Review. *International Journal of Communication Networks and Information Security*, *11*(3).

Farahani, B., Barzegari, M., Aliee, F. S., & Shaik, K. A. (2020). Towards collaborative intelligent IoT eHealth: From device to fog, and cloud. *Microprocessors and Microsystems*, *72*, 102938. doi:10.1016/j.micpro.2019.102938

Hossain, M. S., & Muhammad, G. (2018). Emotion-aware connected healthcare big data towards 5G. *IEEE Internet of Things Journal*, *5*(4), 2399–2406. doi:10.1109/JIOT.2017.2772959

Tiwary, Mahato, Chidar, Chandrol, Shrivastava, & Tripath. (n.d.). Internet of Things (IoT): Research, Architectures and Applications. *International Journal on Future Revolution in Computer Science & Communication Engineering*, *4*(3).

Elijah, O., Rahman, T. A., Orikumhi, I., Leow, C. Y., & Hindia, M. N. (2018). An overview of Internet of Things (IoT) and data analytics in agriculture: Benefits and challenges. *IEEE Internet of Things Journal*, *5*(5), 3758–3773. doi:10.1109/JIOT.2018.2844296

Se-Ra, O. (2017). Security Requirements Analysis for the IoT. Academic Press.

Butt, T., & Afzaal, M. (2019). *Security and Privacy in Smart Cities: Issues and Current Solutions.* doi:10.1007/978-3-030-01659-3_37

Marjani, M., Nasaruddin, F., Gani, A., Karim, A., Hashem, I. A. T., Siddiqa, A., & Yaqoob, I. (2017). Big IoT data analytics: Architecture, opportunities, and open research challenges. *IEEE Access: Practical Innovations, Open Solutions*, 5, 5247–5261. doi:10.1109/ACCESS.2017.2689040

Al-Ali, A. R., Zualkernan, I. A., Rashid, M., Gupta, R., & Alikarar, M. (2017). A smart home energy management system using IoT and big data analytics approach. *IEEE Transactions on Consumer Electronics*, 63(4), 426–434. doi:10.1109/TCE.2017.015014

Farahat, I. S., & Tolba, A. S. (2019). *Security in Smart Cities: Models, Applications, and Challenges*. Springer.

Alahakoon, D., & Yu, X. (2016) Smart electricity meter data intelligence for future energy systems: a survey. *IEEE Trans IndInform, 12*(1), 425–436. .2414355 doi:10.1109/TII.2015

Al-Turjman, Zahmatkesh, & Shehroze. (2019). *An overview of security and privacy in smart cities' IoT communications*. Academic Press.

Barun, B. C. M. (2018). Security and privacy challenges in smart citie. *Sustainable Cities and Society*, 39, 499–507. doi:10.1016/j.scs.2018.02.039

IBM digital analytics. (2020). https://www.ibm.com/in-en/analytics

Alghamdi, T.A. (2020). Energy efficient protocol in wireless sensor network: optimized cluster head selection model. *Telecommun Syst, 74*.

Karagiannis, V., Chatzimisios, P., Vazquez-Gallego, F., & Alonso-Zarate, J. (2015). A survey on application layer protocols for the internet of things. *Transaction on IoT and Cloud Computing, 3*(1), 11-17.

Ngu, Gutierrez, Metsis, Nepal, & Sheng. (2016). IoT Middleware: A Survey on Issues and Enabling Technologies. *IEEE Internet of Things Journal*.

Varga, P., Peto, J., Frankó, A., Balla, D., Haja, D., Janky, F., Soós, G., Ficzere, D., Maliosz, M., & Toka, L. (2020). 5G support for Industrial IoT Applications— Challenges, Solutions, and Research gaps. *Sensors (Basel), 20*(3), 20. doi:10.339020030828 PMID:32033076

Wang, H. Z., Lin, G. W., Wang, J. Q., Gao, W. L., Chen, Y. F., & Duan, Q. L. (2014). Management of Big Data in the Internet of Things in Agriculture Based on Cloud Computing. *Applied Mechanics and Materials, 548*, 1438–1444. doi:10.4028/www.scientific.net/AMM.548-549.1438

Bigquery: Cloud data warehouse. (2020). https://cloud.google.com/bigquery-ml/docs/

Kamble, A., & Bhutad, S. (2018). Survey on Internet of Things (IoT) security issues & solutions. *2nd International Conference on Inventive Systems and Control (ICISC)*, 307-312. 10.1109/ICISC.2018.8399084

Siryani, J., Tanju, B., & Eveleigh, T. J. (2017). A machine learning decision-support system improves the internet of things' smart meter operations. *IEEE Internet of Things Journal, 4*(4), 1056–1066. doi:10.1109/JIOT.2017.2722358

Vashi, S., Ram, J., Modi, J., Verma, S., & Prakash, C. (2017). Internet of Things (IoT): A vision, architectural elements, and security issues. *International Conference on I-SMAC (IoT in Social, Mobile, Analytics and Cloud) (I-SMAC)*, 492-496. 10.1109/I-SMAC.2017.8058399

Hossain, E., Khan, I., Un-Noor, F., Sikander, S. S., & Sunny, M. S. H. (2019). Application of big data and machine learning in smart grid, and associated security concerns: A review. *IEEE Access: Practical Innovations, Open Solutions, 7*, 13960–13988. doi:10.1109/ACCESS.2019.2894819

Internet of Things. (n.d.). In *Wikipedia*. www.wikipedia.com

Li, W., Logenthiran, T., Phan, V. T., & Woo, W. L. (2017). Housing development building management system (hdbms) for optimized electricity bills. *Transactions on Environment and Electrical Engineering*, 2(2), 64–71. doi:10.22149/teee.v2i2.113

Von Hippel, E. (2005). Democratizing innovation: The evolving phenomenon of user innovation. *J. für Betriebswirtschaft*, 55(1), 63–78. doi:10.100711301-004-0002-8

Jacobsson, Boldt, & Carlsson. (2016). A risk analysis of a smart home automation system. *Future Generation Computer Systems, 56*(C), 719 – 733.

OWASP. (n.d.). *Top IoT vulnerabilities - OWASP*. Available: https://www.owasp.org/index.php/Top_IoT_Vulnerabilities

Lin, H., & Bergmann, N. W. (2016). IoT privacy and security challenges for smart home environments. *Information (Basel)*, 7(3), 44. doi:10.3390/info7030044

Scott, D., & Ketel, M. (2016, March). Internet of Things: A useful innovation or security nightmare? In *SoutheastCon 2016* (pp. 1–6). IEEE. doi:10.1109/SECON.2016.7506665

Chaturvedi, A., & Shrivastava, L. (2020). *IoT Based Wireless Sensor Network for Air Pollution Monitoring*. IEEE 9th International Conference on Communication Systems and Network Technologies (CSNT), Gwalior, India.

Iannacci, J. (2018). Internet of things (IoT); internet of everything (IoE); tactile internet; 5G – A (not so evanescent) unifying vision empowered by EH-MEMS (energy harvesting MEMS) and RF-MEMS (radio frequency MEMS). *Sensors and Actuators. A, Physical*, 272, 187–198. Advance online publication. doi:10.1016/j.sna.2018.01.038

Jang-Jaccard, J., & Nepal, S. (2014). A Survey Of Emerging Threats In Cybersecurity. *Journal of Computer and System Sciences*, 80(5), 973–993. doi:10.1016/j.jcss.2014.02.005

Kang, J., Yin, S., & Meng, W. (2014). An Intelligent Storage Management System Based on Cloud Computing and Internet of Things. In *Proceedings of International Conference on Computer Science and Information Technology*. Springer India, 10.1007/978-81-322-1759-6_57

Zschörnig, T., Wehlitz, R., & Franczyk, B. (2020, September). IoT analytics architectures: challenges, solution proposals and future research directions. In *International Conference on Research Challenges in Information Science* (pp. 76-92). Springer. 10.1007/978-3-030-50316-1_5

Werth, O., Guhr, N., & Breitner, M. H. (2020). Smart Home in Private Households: Status Quo, Discussion, and New Insights. *International Journal of Service Science, Management, Engineering, and Technology*, 11(4), 122–136. doi:10.4018/IJSSMET.2020100108

Abomhara, M., & Køien, G. M. (2014, May). Security and privacy in the Internet of Things: Current status and open issues. In 2014 international conference on privacy and security in mobile systems (PRISMS) (pp. 1-8). IEEE.

Perera, C., Ranjan, R., Wang, L., Khan, S. U., & Zomaya, A. Y. (2015). Big Data Privacy in the Internet of Things Era. *IT Professional*, 3(17), 32–39. doi:10.1109/MITP.2015.34

Zaslavsky, A., Perera, C., & Georgakopoulos, D. (2013). *Sensing as a service and big data*. arXiv preprint arXiv:1301.0159.

Misra, P., Rajaraman, V., Dhotrad, K., Warrior, J., & Simmhan, Y. (2015). *An Interoperable Realization of Smart Cities with Plug and Play based Device Management*. arXiv preprint arXiv:1503.00923.

Fouladi, B., & Ghanoun, S. (2013). Security evaluation of the Z-Wave wireless protocol. *Black hat USA, 24*, 1-2.

Rakhmawati, N. A., & Hausenblas, M. (2012). On the Impact of Data Distribution in Federated SPARQL Queries. *Proceedings of 6th IEEE International Conference on Semantic Computing*. 10.1109/ICSC.2012.72

Rakhmawati, N. A., Umbrich, J., Karnstedt, M., Hasnain, A., & Hausenblas, M. (2013). *Querying Over Federated SPARQL Endpoints - A State of the Art Survey*. Technical Report. Digital Enterprise Research Institute. https://arxiv.org/abs/1306.1723

Biswas, A. R., & Giaffreda, R. (2014). IoT and cloud convergence: Opportunities and challenges. *Internet of Things (WF-IoT), IEEE World Forum on.* 10.1109/WF-IoT.2014.6803194

Eisenhauer, M., Vermesan, O., Serrano, M., Guillemin, P., Sundmaeker, H., Tragos, E., Valiño, J., Copigneaux, B., Presser, M., Aagaard, A., Bahr, R., & Darmois, E. (2018). *The Next Generation Internet of Things – Hyperconnectivity and Embedded Intelligence at the Edge*. Academic Press.

Forecast end-user spending on IoT solutions worldwide from 2017 to 2025. (n.d.). https://www.statista.com/statistics/976313/global-iot-market-size

Liang, H., Yang, S., & Li, L. (2019). *Research on routing optimization of WSNs based on improved LEACH protocol. J Wireless Com Network.* doi:10.118613638-019-1509-y

Lopez, J., Roman, R., & Alcaraz, C. (2009). Analysis of Security Threats, Requirements, Technologies And Standards In Wireless Sensor Networks. In Foundations of Security Analysis and Design V. Springer. doi:10.1007/978-3-642-03829-7_10

Abo-Zahhad, M., Farrag, M., Ali, A., & Amin, O. (2015). An Energy consumption model for wireless sensor networks. *IEEE 5th Annual International Conference on Energy Aware Computing Systems and Applications.* 10.1109/ICEAC.2015.7352200

Burrows, M., Abadi, M., & Needham, R. M. (1989). A logic of authentication. *Proceedings of the Royal Society of London A: Mathematical, Physical and Engineering Sciences, 426*(1871), 233–271.

Gebremeskel, G. B., Chai, Y., & Yang, Z. (2014). The Paradigm of Big Data for Augmenting Internet of Vehicle into the Intelligent Cloud Computing Systems. In *Internet of Vehicles–Technologies and Services* (pp. 247–261). Springer International Publishing. doi:10.1007/978-3-319-11167-4_25

Jaloudi, S. (2019). Communication Protocols of an Industrial Internet of Things Environment: A Comparative Study. *Future Internet., 11*(3), 66. Advance online publication. doi:10.3390/fi11030066

Siow, E., Tiropanis, T., & Hall, W. (2018). Analytics for the internet of things: A survey. *ACM Computing Surveys, 51*(4), 1–36. doi:10.1145/3204947

Cirani, S., Picone, M., Gonizzi, P., Veltri, L., & Ferrari, G. (2015). IOT-OAS: An oauth-based authorization service architecture for secure services in iot scenarios. *Sensors Journal, IEEE, 15*(2), 1224–1234. doi:10.1109/JSEN.2014.2361406

European Commission. (2013). *Definition of a research and innovation policy leveraging Cloud Computing and IoT combination*. Tender specifications, SMART 2013/0037.

European Commission. (2015). *Digital Agenda for Europe: The Internet of Things*. https://goo.gl/oNhYOP

Pinto, A. R., Poehls, L. B., Montez, C., & Vargas, F. (2012). *Power optimization for wireless sensor networks*. In *Wireless Sensor Networks - Technology and Applications*. IntechOpen. doi:10.5772/50603

Elahi, H., Munir, K., Eugeni, M., Atek, S., & Gaudenzi, P. (2020). Energy Harvesting towards Self-Powered IoT Devices. *Energies, 13*(21), 5528. doi:10.3390/en13215528

Frustaci, Pace, Aloi, & Fortino. (2018). Evaluating Critical Security Issues of the IoT World: Present and Future Challenges. *IEEE Internet of Things Journal, 5*(4).

International Telecommunication Union. (2012). *Overview of the Internet of Things*. Technical Report. International Telecommunication Union. https://www.itu.int/ITU-T/recommendations/rec.aspx?rec=11559

Lee, K., Murray, D., Hughes, D., & Joosen, W. (2010). Extending sensor networks into the cloud using Amazon web services. *Networked Embedded Systems for Enterprise Applications (NESEA), IEEE International Conference.*

Saha, D., Yousuf, M.R., & Matin, M.A. (2011). Energy efficient scheduling algorithm for S-MAC protocol in wireless sensor network. *International Journal of Wireless and Mobile Networks.*

Jing, Q., Vasilakos, A. V., Wan, J., Lu, J., & Qiu, D. (2014, November). Security of the Internet of Things: Perspectives and challenges. *Wireless Networks, 20*(8), 2481–2501. doi:10.100711276-014-0761-7

Ruiz-Garcia, L., Lunadei, L., Barreiro, P., & Robla, J. I. (2009). A review of wireless sensor technologies and applications in agriculture and food industry: State of the art and current trends. *Sensors (Basel), 9*(6), 4728–4750. doi:10.339090604728 PMID:22408551

Shafiabadi, M. H., Ghafi, A. K., & Manshady, D. D. (2019). *New Method to Improve Energy Savings in Wireless Sensor Networks by Using SOM Neural Network. J Serv Sci Res.* doi:10.100712927-019-0001-x

World Economic Forum. (2012). *The Global Information Technology Report 2012 Living in a Hyperconnected World.* Technical Report. Author.

Abedin, S. F., Alam, M. G. R., Kazmi, S. A., Tran, N. H., Niyato, D., & Hong, C. S. (2018). Resource allocation for ultra-reliable and enhanced mobile broadband IoT applications in fog network. *IEEE Transactions on Communications, 67*(1), 489–502. doi:10.1109/TCOMM.2018.2870888

Alandjani, G. (2018). Features and potential security challenges for IoT enabled devices in smart city environment. *International Journal of Advanced Computer Science and Applications, 9*(8), 231–238. doi:10.14569/IJACSA.2018.090830

Allen, M. (2018). *Building the IoT ecosystem.* https://insightaas.com/building-the-iot-ecosystem/

Asghari, P., Rahmani, A. M., & Javadi, H. H. S. (2019). Internet of Things applications: A systematic review. *Computer Networks, 148*, 241–261. doi:10.1016/j.comnet.2018.12.008

Averian, A. (2018). *A Reference Architecture for Digital Ecosystems, edited by Jaydip Sen, Internet of Things Technology.* In *Applications and Standardization.* IntechOpen. doi:10.5772/intechopen.70907

Barcelo, M. (2016). IoT-Cloud Service Optimization in Next-Generation Smart Environments. Academic Press.

Barnaghi, P., Wang, W., Henson, C., & Taylor, K. (2012). Semantics for the Internet of Things: Early Progress and Back to the Future. *Intel Journal on Semantic Web and Information Systems, 8*(1), 1–21. doi:10.4018/jswis.2012010101

Bauer, M. (2012). *Deliverable D1.4 - Converged architectural reference model for the IoT V2.0.* http://www.iot-a.eu/public/publicdocuments/documents-1/1/1/D1.4/at_download/file

BDO. (n.d.). *Farming in Namibia.* Retrieved November 13, 2020, from https://www.bdo.com.na/en-gb/industries/natural-resources/farming-in-namibia

Bindel, S., Chaumette, S., & Hilt, B. (2015, May). F-ETX: an enhancement of ETX metric for wireless mobile networks. In *International Workshop on Communication Technologies for Vehicles* (pp. 35-46). Springer. 10.1007/978-3-319-17765-6_4

Boulton, C. (2020). *What is digital transformation? A necessary disruption.* https://www.cio.com/article/3211428/what-is-digital-transformation-a-necessary-disruption.html

Canbalaban, E., & Sen, S. (2020, October). A Cross-Layer Intrusion Detection System for RPL-Based Internet of Things. In *International Conference on Ad-Hoc Networks and Wireless* (pp. 214-227). Springer. 10.1007/978-3-030-61746-2_16

Chai, H. S., Choi, J. Y., & Jeong, J. (2015, January). An Enhanced Secure Mobility Management Scheme for Building IoT Applications. In FNC/MobiSPC (pp. 586-591). doi:10.1016/j.procs.2015.07.258

Chen, D., Chang, G., Sun, D., Li, J., Jia, J., & Wang, X. (2011). TRM-IoT: A trust management model based on fuzzy reputation for internet of things. *Computer Science and Information Systems*, 8(4), 1207–1228. doi:10.2298/CSIS110303056C

Chen, R., Guo, J., & Bao, F. (2014). Trust management for SOA-based IoT and its application to service composition. *IEEE Transactions on Services Computing*, 9(3), 482–495. doi:10.1109/TSC.2014.2365797

Csáki, C. (2019). *Open Data Ecosystems: A Comparison of Visual Models. In Electronic Government and the Information Systems Perspective*. Springer. doi:10.1007/978-3-030-27523-5_2

Cubarrubia, A., & Perry, P. (2016). *Creating a Thriving Postsecondary Education Data Ecosystem*. http://www.ihep.org/research/publications/creating-thriving-postsecondary-education-data-ecosystem

Curry, E., & Ojo, A. (2020). Enabling Knowledge Flows in an Intelligent Systems Data Ecosystem, Real-time Linked Dataspaces. doi:10.1007/978-3-030-29665-0_2

Curry, E., & Sheth, A. (2018). Next-Generation Smart Environments: From System of Systems to Data Ecosystems. *IEEE Intelligent Systems*, ●●●, 69–75.

Curry, E., & Sheth, A. (2018). Next-Generation Smart Environments: From System of Systems to Data Ecosystems. *IEEE Intelligent Systems*, 33(3), 69–76. doi:10.1109/MIS.2018.033001418

Da Xu, L., He, W., & Li, S. (2014). Internet of things in industries: A survey. *IEEE Transactions on Industrial Informatics*, 10(4), 2233–2243. doi:10.1109/TII.2014.2300753

Dai, M., Su, Z., Li, R., Wang, Y., Ni, J., & Fang, D. (2020). An Edge-Driven Security Framework for Intelligent Internet of Things. *IEEE Network*, 34(5), 39–45. Advance online publication. doi:10.1109/MNET.011.2000068

Das, S., Ghosh, I., Banerjee, G., & Sarkar, U. (2018). *Artificial Intelligence in Agriculture: A Literature Survey*. Academic Press.

de la Boutetière, H., Montagner, A., & Reich, A. (2018). *Unlocking success in digital transformations*. https://www.mckinsey.com/business-functions/organization/our-insights/unlocking-success-in-digital-transformations

Demchenko, Y., Laat de, C., & Membrey, P. (2014). Defining architecture components of the Big Data Ecosystem. *Intel. Conf. on Collaboration Technologies and Systems (CTS)*, 104-112. . doi:10.1109/CTS.2014.6867550

Deng, H. (2016). Big data ecosystem model, and application in the city. *Journal of Big Data Research*, 2(2), 68–75.

Digital Transformation. (n.d.). In *Wikipedia*. https://en.wikipedia.org/wiki/Digital_transformation

Digiteum. (2020). *How IoT is used in Education: IoT Applications in Education*. https://www.digiteum.com/iot-applications-education

Dua, A. (2018). *Smart education is more than just Advanced Learning Methods*. https://yourstory.com/2018/05/smart-education-advanced-learning

Faustine, A., Mvuma, A. N., Mongi, H. J., Gabriel, M. C., Tenge, A. J., & Kucel, S. B. (2014). *Wireless Sensor Networks for Water Quality Monitoring and Control within Lake Victoria Basin: Prototype Development*. Wireless Sensor Network. doi:10.4236/wsn.2014.612027

Ferrer, T., Céspedes, S., & Becerra, A. (2019). Review and evaluation of MAC protocols for satellite IoT systems using nanosatellites. *Sensors (Basel)*, *19*(8), 1947. doi:10.339019081947 PMID:31027250

Frost, S. (n.d.). *Mega Trends: Smart is the New Green.* https://www.frost.com/prod/servlet/our-services-age.pag?mode=open&sid=230169625

Goap, A., Sharma, D., Shukla, A. K., & Rama Krishna, C. (2018). An IoT based smart irrigation management system using Machine learning and open source technologies. *Computers and Electronics in Agriculture*, *155*, 41–49. doi:10.1016/j.compag.2018.09.040

Gomes, J., Rodrigues, J. J., Rabêlo, R. A., Kumar, N., & Kozlov, S. (2019). IoT-Enabled Gas Sensors: Technologies, Applications, and Opportunities. *Journal of Sensor and Actuator Networks*, *8*(4), 57. doi:10.3390/jsan8040057

Gutierrez, J., Villa-Medina, J. F., Nieto-Garibay, A., & Porta-Gandara, M. A. (2014). Automated Irrigation System Using a Wireless Sensor Network and GPRS Module. *IEEE Transactions on Instrumentation and Measurement*, *63*(1), 166–176. doi:10.1109/TIM.2013.2276487

Hannon, V., Patton, A., & Temperley, J. (2011). *Developing an Innovation Ecosystem for Education.* White Paper, Cisco Innovation Unit.

Hashemi, S. Y., & Aliee, F. S. (2019). Dynamic and comprehensive trust model for IoT and its integration into RPL. *The Journal of Supercomputing*, *75*(7), 3555–3584. doi:10.100711227-018-2700-3

Hashiyana, V., Suresh, N., Haiduwa, T., Mbewe, D., & Ujakpa, M. M. (2020). Co-design of an Agricultural Management Application for Small-Scale Farmers. *2020 IST-Africa Conference (IST-Africa)*, 1–7.

Hellaoui, H., Bouabdallah, A., & Koudil, M. (2016, November). Tas-iot: trust-based adaptive security in the iot. In *2016 IEEE 41st Conference on Local Computer Networks (LCN)* (pp. 599-602). IEEE. 10.1109/LCN.2016.101

Hoel, T., & Mason, J. (2018). Standards for smart education – towards a development framework. *Smart Learn. Environ.*, *5*(1), 3. doi:10.118640561-018-0052-3

Infotech. (2020). *Smart Classroom Technology-IoT in Education Industry.* https://medium.com/@chapter247infotech

I-Scoop. (2020). *Digital transformation: online guide to digital business transformation.* https://www.i-scoop.eu/digital-transformation/

Jain, R., & Kashyap, I. (2019). An QoS aware link defined OLSR (LD-OLSR) routing protocol for MANETs. *Wireless Personal Communications*, *108*(3), 1745–1758. doi:10.100711277-019-06494-9

Jain, R., & Kashyap, I. (2019). Performance Evaluation of OLSR-MD Routing Protocol for MANETS. In *Advances in Computer Communication and Computational Sciences* (pp. 101–108). Springer. doi:10.1007/978-981-13-6861-5_9

Jain, R., & Kashyap, I. (2020). Energy-Based Improved MPR Selection in OLSR Routing Protocol. In *Data Management, Analytics and Innovation* (pp. 583–599). Springer. doi:10.1007/978-981-32-9949-8_41

Jamali, S., Fotohi, R., & Analoui, M. (2018). An artificial immune system based method for defense against wormhole attack in mobile adhoc networks. *Tabriz Journal of Electrical Engineering*, *47*(4), 1407–1419.

Jason, N. (2014). Smart Learning for the Next Generation Education Environment. In *2014 International Conference on Intelligent Environments*. IEEE Xplore. 10.1109/IE.2014.73

Jeong, J.-S., Kim, M., & Yoo, K.-H. (2013). A Content Oriented Smart Education System based on Cloud Computing. *Intel. J of Multimedia and Ubiquitous Engineering*, *8*(6), 313–328. doi:10.14257/ijmue.2013.8.6.31

Kamble, A., Malemath, V. S., & Patil, D. (2017, February). Security attacks and secure routing protocols in RPL-based Internet of Things: Survey. In *2017 International Conference on Emerging Trends & Innovation in ICT (ICEI)* (pp. 33-39). IEEE. 10.1109/ETIICT.2017.7977006

Kitsios, F., Papachristos, N., & Kamariotou, M. (2017). Business Models for Open Data Ecosystem: Challenges and Motivations for Entrepreneurship and Innovation. *IEEE 19th Conf. on Business Informatics (CBI)*, 398-407. 10.1109/CBI.2017.51

Lamaazi, H., & Benamar, N. (2019). A novel approach for RPL assessment based on the objective function and trickle optimizations. *Wireless Communications and Mobile Computing*, *2019*, 2019. doi:10.1155/2019/4605095

Lee, R. S. T. (2020). Smart Education. In *Artificial Intelligence in Daily Life*. Springer. doi:10.1007/978-981-15-7695-9_11

Liehr, S., Brenda, M., Cornel, P., Deffner, J., Felmeden, J., Jokisch, A., Kluge, T., Müller, K., Röhrig, J., Stibitz, V., & Urban, W. (2016). From the Concept to the Tap—Integrated Water Resources Management in Northern Namibia. In D. Borchardt, J. J. Bogardi, & R. B. Ibisch (Eds.), *Integrated Water Resources Management: Concept, Research and Implementation* (pp. 683–717). Springer International Publishing. doi:10.1007/978-3-319-25071-7_26

Liehr, S., Kramm, J., Jokisch, A., & Müller, K. (2018). Integrated Water Resources Management in Water-scarce Regions: Water Harvesting, Groundwater Desalination and Water Reuse in Namibia. *Water Intelligence Online*, *17*, 9781780407913. doi:10.2166/9781780407913

Li, S., Da Xu, L., & Zhao, S. (2015). The internet of things: A survey. *Information Systems Frontiers*, *17*(2), 243–259. doi:10.100710796-014-9492-7

Li, S., Da Xu, L., & Zhao, S. (2018). 5G Internet of Things: A survey. *Journal of Industrial Information Integration*, *10*, 1–9. doi:10.1016/j.jii.2018.01.005

Lu, H., Li, Y., Chen, M., Kim, H., & Serikawa, S. (2018). Brain Intelligence: Go beyond Artificial Intelligence. *Mobile Networks and Applications*, *23*(2), 368–375. doi:10.100711036-017-0932-8

Melis, A., Prandini, M., Sartori, L., & Callegati, F. (2016, September). Public transportation, IoT, trust and urban habits. In *International conference on internet science* (pp. 318-325). Springer. 10.1007/978-3-319-45982-0_27

Mérida García, A., Fernández García, I., Camacho Poyato, E., Montesinos Barrios, P., & Rodríguez Díaz, J. A. (2018). Coupling irrigation scheduling with solar energy production in a smart irrigation management system. *Journal of Cleaner Production*, *175*, 670–682. doi:10.1016/j.jclepro.2017.12.093

MixPanel. (2020). *How to create a Data Ecosystem*. https://mixpanel.com/topics/what-is-a-data-ecosystem/

Mohanta, B. K., Jena, D., Satapathy, U., & Patnaik, S. (2020). Survey on IoT Security: Challenges and Solution using Machine Learning, Artificial Intelligence and Blockchain Technology. *Internet of Things*, 100227.

Munyayi, B. S. (2015). Water Innovation: innovativee approaches towards water security in Namibia. Drfn.

Niu, Y., Zhang, J., Wang, A., & Chen, C. (2019). An efficient collision power attack on AES encryption in edge computing. *IEEE Access: Practical Innovations, Open Solutions*, *7*, 18734–18748. doi:10.1109/ACCESS.2019.2896256

Oliveira, L., Rodrigues, J. J., Kozlov, S. A., Rabêlo, R. A., & Albuquerque, V. H. C. D. (2019). MAC layer protocols for Internet of Things: A survey. *Future Internet*, *11*(1), 16. doi:10.3390/fi11010016

Oliveira, S., Barros Lima, G. D. F., & Farias Lóscio, B. (2019). Investigations into Data Ecosystems: A systematic mapping study. *Knowledge and Information Systems*, *61*, 589–630. doi:10.100710115-018-1323-6

Oriel, A. (2020). *Building a Strong Data Ecosystem for AI-Powered Organizations.* https://www.analyticsinsight.net/building-strong-data-ecosystem-ai-powered-organizations/

Parthasarathy, S., Tung, T., Munnelly, S., & Joshi, S. K. (2019). *Bringing Data Together: A Modern Data Ecosystem.* Accenture. https://www.accenture.com

Pu, C. (2019, February). Spam dis attack against routing protocol in the internet of things. In *2019 International Conference on Computing, Networking and Communications (ICNC)* (pp. 73-77). IEEE. 10.1109/ICCNC.2019.8685628

Putra, R. R. J., & Putro, B. L. (2019). *J. Phys.: Conf. Ser. 1280 032029.* IOP Publishing. doi:10.1088/1742-6596/1280/3/032029

Qin, T., Wang, B., Chen, R., Qin, Z., & Wang, L. (2019). IMLADS: Intelligent maintenance and lightweight anomaly detection system for internet of things. *Sensors (Basel)*, *19*(4), 958. doi:10.339019040958 PMID:30813486

Ramachandran, V., Ramalakshmi, R., & Srinivasan, S. (2018). An Automated Irrigation System for Smart Agriculture Using the Internet of Things. *2018 15th International Conference on Control, Automation, Robotics and Vision (ICARCV)*, 210–215. 10.1109/ICARCV.2018.8581221

Raman, R. (2020). *IoT and its impact on education.* https://www.deccanherald.com/supplements/dh-education/iot-and-its-impact-on-education-845493.html

Ranathunga, T., Marfievici, R., McGibney, A., & Rea, S. (2020, June). A DLT-based Trust Framework for IoT Ecosystems. In *2020 International Conference on Cyber Security and Protection of Digital Services (Cyber Security)* (pp. 1-8). IEEE.

Ray P P (2016). A survey on Internet of Things architectures. *Journal of King Saud University – Computer and Information Sciences*, 1319-1578. . doi:10.1016/j.jksuci.2016.10.003

RedHat. (2020). *What is digital transformation?* https://www.redhat.com/en/topics/digital-transformation/what-is-digital-transformation

Roy, M. (2020). AI Intervention in Education Systems of India: An Analysis. *Solid State Technology*, *63*(2), 1395–1402.

Ruan, Y., Durresi, A., & Uslu, S. (2018, May). Trust assessment for internet of things in multi-access edge computing. In *2018 IEEE 32nd International Conference on Advanced Information Networking and Applications (AINA)* (pp. 1155-1161). IEEE. 10.1109/AINA.2018.00165

Sanmartin, P., Jabba, D., Sierra, R., & Martinez, E. (2018). Objective function BF-ETX for RPL routing protocol. *IEEE Latin America Transactions*, *16*(8), 2275–2281. doi:10.1109/TLA.2018.8528246

Shikangalah, R. N., & Mapani, B. S. (2020). A review of bush encroachment in Namibia: From a problem to an opportunity? *Journal of Rangeland Science*, *10*(3), 251–266.

Smart Systems and Internet of Things Platforms. (n.d.). *Overview of Research and Analysis and Summary Findings, Smart System Design.* Barbor Research. https://niolabs.com/app/uploads/2017/10/HRI_Platform-Rpt-Summary_12-October-2017.pdf

Smith, M. J. (2020). Getting value from artificial intelligence in agriculture. *Animal Production Science*, *60*(1), 46. doi:10.1071/AN18522

Somses, S., Bopape, M.-J. M., Ndarana, T., Fridlind, A., Matsui, T., Phaduli, E., Limbo, A., Maikhudumu, S., Maisha, R., & Rakate, E. (2020). Convection Parametrization and Multi-Nesting Dependence of a Heavy Rainfall Event over Namibia with Weather Research and Forecasting (WRF) Model. *Climate (Basel)*, *8*(10), 112. doi:10.3390/cli8100112

Subramanian & Srivastava. (2017). *Architecture Patterns for the Next-generation Data Ecosystem.* Tata Consulting Services.

Subramanian, N., GB, S. M., Martin, J. P., & Chandrasekaran, K. (2020, January). HTmRPL++: A Trust-Aware RPL Routing Protocol for Fog Enabled Internet of Things. In *2020 International Conference on COMmunicationSystems & NETworkS (COMSNETS)* (pp. 1-5). IEEE. 10.1109/COMSNETS48256.2020.9027387

Suresh, N., Hashiyana, V., Kulula, V. P., & Thotappa, S. (2019). Smart Water Level Monitoring System for Farmers. In D. Goyal, S. Balamurugan, S.-L. Peng, & D. S. Jat (Eds.), *The IoT and the Next Revolutions Automating the World* (pp. 213–228). IGI Global. doi:10.4018/978-1-5225-9246-4.ch014

Tassey, M., Gray, E., & Cottrell, S. (2020). *Data Transfer in the Larger Education Ecosystem.* United States Department of Education, Privacy Technical Assistance Center. https://studentprivacy.ed.gov/sites/default/files/resource_document/file/DataTransfer-in-the-Larger-Education-Ecosystem.pdf

Ulmer, J., Belaud, J., & Le Lann, J. (2013). A pivotal-based approach for enterprise business process and IS integration. *Enterprise Information Systems*, 7(1), 61–78. doi:10.1080/17517575.2012.700326

UNDP. (2016). *Data Ecosystems for Sustainable Development, an Assessment of Six Pilot Countries.* Report, United Nations Development Programme.

Vaitsis, C., Hervatis, V., & Zary, N. (2016). Introduction to Big Data in Education and Its Contribution to the Quality Improvement Processes. In. Big Data on Real-World Applications. IntechOpen Science. doi:10.5772/63896

Veena, D., Ayush, A., Chandan, K., Raunak, R., & Rochak, B. (2013). A Real time implementation of a GSM based Automated Irrigation Control System using Drip Irrigation Methology. *International Journal of Scientific and Engineering Research*, 4(5), 146–151.

Vermesan, O., & Friess, P. (2013). *Internet of Things: Converging Technologies for Smart Environments and Integrated Ecosystems.* River Publishers Series in Communications.

Whitmore, A., Agarwal, A., & Da Xu, L. (2015). The Internet of Things—A survey of topics and trends. *Information Systems Frontiers*, 17(2), 261–274. doi:10.100710796-014-9489-2

Wu, D., Shi, H., Wang, H., Wang, R., & Fang, H. (2019). A Feature-Based Learning System for Internet of Things Applications. *IEEE Internet of Things Journal*, 6(2), 1928–1937. doi:10.1109/JIOT.2018.2884485

Yang, D. L., Liu, F., & Liang, Y. D. (2010, December). A survey of the internet of things. In *Proceedings of the 1st International Conference on E-Business Intelligence (ICEBI2010)*. Atlantis Press. 10.2991/icebi.2010.72

Yang, Y., Wu, L., Yin, G., Li, L., & Zhao, H. (2017). A survey on security and privacy issues in Internet-of-Things. *IEEE Internet of Things Journal*, 4(5), 1250–1258. doi:10.1109/JIOT.2017.2694844

Yan, Z., Zhang, P., & Vasilakos, A. V. (2014). A survey on trust management for Internet of Things. *Journal of Network and Computer Applications*, 42, 120–134. doi:10.1016/j.jnca.2014.01.014

Yavuz, F. Y., Devrim, Ü. N. A. L., & Ensar, G. Ü. L. (2018). Deep learning for detection of routing attacks in the internet of things. *International Journal of Computational Intelligence Systems*, 12(1), 39–58. doi:10.2991/ijcis.2018.25905181

Yildiz, M. (2017). *Introduction to IoT Ecosystem, a Technical, Architectural & Solution.* https://medium.com/illumination-curated/introduction-to-iot-ecosystem-25b359c8cf23

Zhao, W., Lin, S., Han, J., Xu, R., & Hou, L. (2017). Design and Implementation of Smart Irrigation System Based on LoRa. *2017 IEEE Globecom Workshops (GC Wkshps)*, 1–6. doi:10.1109/GLOCOMW.2017.8269115

Related References

To continue our tradition of advancing information science and technology research, we have compiled a list of recommended IGI Global readings. These references will provide additional information and guidance to further enrich your knowledge and assist you with your own research and future publications.

Aasi, P., Rusu, L., & Vieru, D. (2017). The Role of Culture in IT Governance Five Focus Areas: A Literature Review. *International Journal of IT/Business Alignment and Governance, 8*(2), 42-61. doi:10.4018/IJITBAG.2017070103

Abdrabo, A. A. (2018). Egypt's Knowledge-Based Development: Opportunities, Challenges, and Future Possibilities. In A. Alraouf (Ed.), *Knowledge-Based Urban Development in the Middle East* (pp. 80–101). Hershey, PA: IGI Global. doi:10.4018/978-1-5225-3734-2.ch005

Abu Doush, I., & Alhami, I. (2018). Evaluating the Accessibility of Computer Laboratories, Libraries, and Websites in Jordanian Universities and Colleges. *International Journal of Information Systems and Social Change, 9*(2), 44–60. doi:10.4018/IJISSC.2018040104

Adeboye, A. (2016). Perceived Use and Acceptance of Cloud Enterprise Resource Planning (ERP) Implementation in the Manufacturing Industries. *International Journal of Strategic Information Technology and Applications, 7*(3), 24–40. doi:10.4018/IJSITA.2016070102

Adegbore, A. M., Quadri, M. O., & Oyewo, O. R. (2018). A Theoretical Approach to the Adoption of Electronic Resource Management Systems (ERMS) in Nigerian University Libraries. In A. Tella & T. Kwanya (Eds.), *Handbook of Research on Managing Intellectual Property in Digital Libraries* (pp. 292–311). Hershey, PA: IGI Global. doi:10.4018/978-1-5225-3093-0.ch015

Adhikari, M., & Roy, D. (2016). Green Computing. In G. Deka, G. Siddesh, K. Srinivasa, & L. Patnaik (Eds.), *Emerging Research Surrounding Power Consumption and Performance Issues in Utility Computing* (pp. 84–108). Hershey, PA: IGI Global. doi:10.4018/978-1-4666-8853-7.ch005

Afolabi, O. A. (2018). Myths and Challenges of Building an Effective Digital Library in Developing Nations: An African Perspective. In A. Tella & T. Kwanya (Eds.), *Handbook of Research on Managing Intellectual Property in Digital Libraries* (pp. 51–79). Hershey, PA: IGI Global. doi:10.4018/978-1-5225-3093-0.ch004

Agarwal, R., Singh, A., & Sen, S. (2016). Role of Molecular Docking in Computer-Aided Drug Design and Development. In S. Dastmalchi, M. Hamzeh-Mivehroud, & B. Sokouti (Eds.), *Applied Case Studies and Solutions in Molecular Docking-Based Drug Design* (pp. 1–28). Hershey, PA: IGI Global. doi:10.4018/978-1-5225-0362-0.ch001

Ali, O., & Soar, J. (2016). Technology Innovation Adoption Theories. In L. Al-Hakim, X. Wu, A. Koronios, & Y. Shou (Eds.), *Handbook of Research on Driving Competitive Advantage through Sustainable, Lean, and Disruptive Innovation* (pp. 1–38). Hershey, PA: IGI Global. doi:10.4018/978-1-5225-0135-0.ch001

Alsharo, M. (2017). Attitudes Towards Cloud Computing Adoption in Emerging Economies. *International Journal of Cloud Applications and Computing*, 7(3), 44–58. doi:10.4018/IJCAC.2017070102

Amer, T. S., & Johnson, T. L. (2016). Information Technology Progress Indicators: Temporal Expectancy, User Preference, and the Perception of Process Duration. *International Journal of Technology and Human Interaction*, 12(4), 1–14. doi:10.4018/IJTHI.2016100101

Amer, T. S., & Johnson, T. L. (2017). Information Technology Progress Indicators: Research Employing Psychological Frameworks. In A. Mesquita (Ed.), *Research Paradigms and Contemporary Perspectives on Human-Technology Interaction* (pp. 168–186). Hershey, PA: IGI Global. doi:10.4018/978-1-5225-1868-6.ch008

Anchugam, C. V., & Thangadurai, K. (2016). Introduction to Network Security. In D. G., M. Singh, & M. Jayanthi (Eds.), Network Security Attacks and Countermeasures (pp. 1-48). Hershey, PA: IGI Global. doi:10.4018/978-1-4666-8761-5.ch001

Anchugam, C. V., & Thangadurai, K. (2016). Classification of Network Attacks and Countermeasures of Different Attacks. In D. G., M. Singh, & M. Jayanthi (Eds.), Network Security Attacks and Countermeasures (pp. 115-156). Hershey, PA: IGI Global. doi:10.4018/978-1-4666-8761-5.ch004

Anohah, E. (2016). Pedagogy and Design of Online Learning Environment in Computer Science Education for High Schools. *International Journal of Online Pedagogy and Course Design*, 6(3), 39–51. doi:10.4018/IJOPCD.2016070104

Anohah, E. (2017). Paradigm and Architecture of Computing Augmented Learning Management System for Computer Science Education. *International Journal of Online Pedagogy and Course Design*, 7(2), 60–70. doi:10.4018/IJOPCD.2017040105

Anohah, E., & Suhonen, J. (2017). Trends of Mobile Learning in Computing Education from 2006 to 2014: A Systematic Review of Research Publications. *International Journal of Mobile and Blended Learning*, 9(1), 16–33. doi:10.4018/IJMBL.2017010102

Assis-Hassid, S., Heart, T., Reychav, I., & Pliskin, J. S. (2016). Modelling Factors Affecting Patient-Doctor-Computer Communication in Primary Care. *International Journal of Reliable and Quality E-Healthcare*, 5(1), 1–17. doi:10.4018/IJRQEH.2016010101

Bailey, E. K. (2017). Applying Learning Theories to Computer Technology Supported Instruction. In M. Grassetti & S. Brookby (Eds.), *Advancing Next-Generation Teacher Education through Digital Tools and Applications* (pp. 61–81). Hershey, PA: IGI Global. doi:10.4018/978-1-5225-0965-3.ch004

Balasubramanian, K. (2016). Attacks on Online Banking and Commerce. In K. Balasubramanian, K. Mala, & M. Rajakani (Eds.), *Cryptographic Solutions for Secure Online Banking and Commerce* (pp. 1–19). Hershey, PA: IGI Global. doi:10.4018/978-1-5225-0273-9.ch001

Baldwin, S., Opoku-Agyemang, K., & Roy, D. (2016). Games People Play: A Trilateral Collaboration Researching Computer Gaming across Cultures. In K. Valentine & L. Jensen (Eds.), *Examining the Evolution of Gaming and Its Impact on Social, Cultural, and Political Perspectives* (pp. 364–376). Hershey, PA: IGI Global. doi:10.4018/978-1-5225-0261-6.ch017

Banerjee, S., Sing, T. Y., Chowdhury, A. R., & Anwar, H. (2018). Let's Go Green: Towards a Taxonomy of Green Computing Enablers for Business Sustainability. In M. Khosrow-Pour (Ed.), *Green Computing Strategies for Competitive Advantage and Business Sustainability* (pp. 89–109). Hershey, PA: IGI Global. doi:10.4018/978-1-5225-5017-4.ch005

Basham, R. (2018). Information Science and Technology in Crisis Response and Management. In M. Khosrow-Pour, D.B.A. (Ed.), Encyclopedia of Information Science and Technology, Fourth Edition (pp. 1407-1418). Hershey, PA: IGI Global. doi:10.4018/978-1-5225-2255-3.ch121

Batyashe, T., & Iyamu, T. (2018). Architectural Framework for the Implementation of Information Technology Governance in Organisations. In M. Khosrow-Pour, D.B.A. (Ed.), Encyclopedia of Information Science and Technology, Fourth Edition (pp. 810-819). Hershey, PA: IGI Global. doi:10.4018/978-1-5225-2255-3.ch070

Bekleyen, N., & Çelik, S. (2017). Attitudes of Adult EFL Learners towards Preparing for a Language Test via CALL. In D. Tafazoli & M. Romero (Eds.), *Multiculturalism and Technology-Enhanced Language Learning* (pp. 214–229). Hershey, PA: IGI Global. doi:10.4018/978-1-5225-1882-2.ch013

Bennett, A., Eglash, R., Lachney, M., & Babbitt, W. (2016). Design Agency: Diversifying Computer Science at the Intersections of Creativity and Culture. In M. Raisinghani (Ed.), *Revolutionizing Education through Web-Based Instruction* (pp. 35–56). Hershey, PA: IGI Global. doi:10.4018/978-1-4666-9932-8.ch003

Bergeron, F., Croteau, A., Uwizeyemungu, S., & Raymond, L. (2017). A Framework for Research on Information Technology Governance in SMEs. In S. De Haes & W. Van Grembergen (Eds.), *Strategic IT Governance and Alignment in Business Settings* (pp. 53–81). Hershey, PA: IGI Global. doi:10.4018/978-1-5225-0861-8.ch003

Bhatt, G. D., Wang, Z., & Rodger, J. A. (2017). Information Systems Capabilities and Their Effects on Competitive Advantages: A Study of Chinese Companies. *Information Resources Management Journal*, *30*(3), 41–57. doi:10.4018/IRMJ.2017070103

Bogdanoski, M., Stoilkovski, M., & Risteski, A. (2016). Novel First Responder Digital Forensics Tool as a Support to Law Enforcement. In M. Hadji-Janev & M. Bogdanoski (Eds.), *Handbook of Research on Civil Society and National Security in the Era of Cyber Warfare* (pp. 352–376). Hershey, PA: IGI Global. doi:10.4018/978-1-4666-8793-6.ch016

Boontarig, W., Papasratorn, B., & Chutimaskul, W. (2016). The Unified Model for Acceptance and Use of Health Information on Online Social Networks: Evidence from Thailand. *International Journal of E-Health and Medical Communications*, *7*(1), 31–47. doi:10.4018/IJEHMC.2016010102

Brown, S., & Yuan, X. (2016). Techniques for Retaining Computer Science Students at Historical Black Colleges and Universities. In C. Prince & R. Ford (Eds.), *Setting a New Agenda for Student Engagement and Retention in Historically Black Colleges and Universities* (pp. 251–268). Hershey, PA: IGI Global. doi:10.4018/978-1-5225-0308-8.ch014

Burcoff, A., & Shamir, L. (2017). Computer Analysis of Pablo Picasso's Artistic Style. *International Journal of Art, Culture and Design Technologies*, *6*(1), 1–18. doi:10.4018/IJACDT.2017010101

Byker, E. J. (2017). I Play I Learn: Introducing Technological Play Theory. In C. Martin & D. Polly (Eds.), *Handbook of Research on Teacher Education and Professional Development* (pp. 297–306). Hershey, PA: IGI Global. doi:10.4018/978-1-5225-1067-3.ch016

Calongne, C. M., Stricker, A. G., Truman, B., & Arenas, F. J. (2017). Cognitive Apprenticeship and Computer Science Education in Cyberspace: Reimagining the Past. In A. Stricker, C. Calongne, B. Truman, & F. Arenas (Eds.), *Integrating an Awareness of Selfhood and Society into Virtual Learning* (pp. 180–197). Hershey, PA: IGI Global. doi:10.4018/978-1-5225-2182-2.ch013

Carlton, E. L., Holsinger, J. W. Jr, & Anunobi, N. (2016). Physician Engagement with Health Information Technology: Implications for Practice and Professionalism. *International Journal of Computers in Clinical Practice*, *1*(2), 51–73. doi:10.4018/IJCCP.2016070103

Carneiro, A. D. (2017). Defending Information Networks in Cyberspace: Some Notes on Security Needs. In M. Dawson, D. Kisku, P. Gupta, J. Sing, & W. Li (Eds.), Developing Next-Generation Countermeasures for Homeland Security Threat Prevention (pp. 354-375). Hershey, PA: IGI Global. doi:10.4018/978-1-5225-0703-1.ch016

Cavalcanti, J. C. (2016). The New "ABC" of ICTs (Analytics + Big Data + Cloud Computing): A Complex Trade-Off between IT and CT Costs. In J. Martins & A. Molnar (Eds.), *Handbook of Research on Innovations in Information Retrieval, Analysis, and Management* (pp. 152–186). Hershey, PA: IGI Global. doi:10.4018/978-1-4666-8833-9.ch006

Chase, J. P., & Yan, Z. (2017). Affect in Statistics Cognition. In *Assessing and Measuring Statistics Cognition in Higher Education Online Environments: Emerging Research and Opportunities* (pp. 144–187). Hershey, PA: IGI Global. doi:10.4018/978-1-5225-2420-5.ch005

Chen, C. (2016). Effective Learning Strategies for the 21st Century: Implications for the E-Learning. In M. Anderson & C. Gavan (Eds.), *Developing Effective Educational Experiences through Learning Analytics* (pp. 143–169). Hershey, PA: IGI Global. doi:10.4018/978-1-4666-9983-0.ch006

Chen, E. T. (2016). Examining the Influence of Information Technology on Modern Health Care. In P. Manolitzas, E. Grigoroudis, N. Matsatsinis, & D. Yannacopoulos (Eds.), *Effective Methods for Modern Healthcare Service Quality and Evaluation* (pp. 110–136). Hershey, PA: IGI Global. doi:10.4018/978-1-4666-9961-8.ch006

Cimermanova, I. (2017). Computer-Assisted Learning in Slovakia. In D. Tafazoli & M. Romero (Eds.), *Multiculturalism and Technology-Enhanced Language Learning* (pp. 252–270). Hershey, PA: IGI Global. doi:10.4018/978-1-5225-1882-2.ch015

Cipolla-Ficarra, F. V., & Cipolla-Ficarra, M. (2018). Computer Animation for Ingenious Revival. In F. Cipolla-Ficarra, M. Ficarra, M. Cipolla-Ficarra, A. Quiroga, J. Alma, & J. Carré (Eds.), *Technology-Enhanced Human Interaction in Modern Society* (pp. 159–181). Hershey, PA: IGI Global. doi:10.4018/978-1-5225-3437-2.ch008

Cockrell, S., Damron, T. S., Melton, A. M., & Smith, A. D. (2018). Offshoring IT. In M. Khosrow-Pour, D.B.A. (Ed.), Encyclopedia of Information Science and Technology, Fourth Edition (pp. 5476-5489). Hershey, PA: IGI Global. doi:10.4018/978-1-5225-2255-3.ch476

Coffey, J. W. (2018). Logic and Proof in Computer Science: Categories and Limits of Proof Techniques. In J. Horne (Ed.), *Philosophical Perceptions on Logic and Order* (pp. 218–240). Hershey, PA: IGI Global. doi:10.4018/978-1-5225-2443-4.ch007

Dale, M. (2017). Re-Thinking the Challenges of Enterprise Architecture Implementation. In M. Tavana (Ed.), *Enterprise Information Systems and the Digitalization of Business Functions* (pp. 205–221). Hershey, PA: IGI Global. doi:10.4018/978-1-5225-2382-6.ch009

Das, A., Dasgupta, R., & Bagchi, A. (2016). Overview of Cellular Computing-Basic Principles and Applications. In J. Mandal, S. Mukhopadhyay, & T. Pal (Eds.), *Handbook of Research on Natural Computing for Optimization Problems* (pp. 637–662). Hershey, PA: IGI Global. doi:10.4018/978-1-5225-0058-2.ch026

De Maere, K., De Haes, S., & von Kutzschenbach, M. (2017). CIO Perspectives on Organizational Learning within the Context of IT Governance. *International Journal of IT/Business Alignment and Governance, 8*(1), 32-47. doi:10.4018/IJITBAG.2017010103

Demir, K., Çaka, C., Yaman, N. D., İslamoğlu, H., & Kuzu, A. (2018). Examining the Current Definitions of Computational Thinking. In H. Ozcinar, G. Wong, & H. Ozturk (Eds.), *Teaching Computational Thinking in Primary Education* (pp. 36–64). Hershey, PA: IGI Global. doi:10.4018/978-1-5225-3200-2.ch003

Deng, X., Hung, Y., & Lin, C. D. (2017). Design and Analysis of Computer Experiments. In S. Saha, A. Mandal, A. Narasimhamurthy, S. V, & S. Sangam (Eds.), Handbook of Research on Applied Cybernetics and Systems Science (pp. 264-279). Hershey, PA: IGI Global. doi:10.4018/978-1-5225-2498-4.ch013

Denner, J., Martinez, J., & Thiry, H. (2017). Strategies for Engaging Hispanic/Latino Youth in the US in Computer Science. In Y. Rankin & J. Thomas (Eds.), *Moving Students of Color from Consumers to Producers of Technology* (pp. 24–48). Hershey, PA: IGI Global. doi:10.4018/978-1-5225-2005-4.ch002

Devi, A. (2017). Cyber Crime and Cyber Security: A Quick Glance. In R. Kumar, P. Pattnaik, & P. Pandey (Eds.), *Detecting and Mitigating Robotic Cyber Security Risks* (pp. 160–171). Hershey, PA: IGI Global. doi:10.4018/978-1-5225-2154-9.ch011

Dores, A. R., Barbosa, F., Guerreiro, S., Almeida, I., & Carvalho, I. P. (2016). Computer-Based Neuropsychological Rehabilitation: Virtual Reality and Serious Games. In M. Cruz-Cunha, I. Miranda, R. Martinho, & R. Rijo (Eds.), *Encyclopedia of E-Health and Telemedicine* (pp. 473–485). Hershey, PA: IGI Global. doi:10.4018/978-1-4666-9978-6.ch037

Doshi, N., & Schaefer, G. (2016). Computer-Aided Analysis of Nailfold Capillaroscopy Images. In D. Fotiadis (Ed.), *Handbook of Research on Trends in the Diagnosis and Treatment of Chronic Conditions* (pp. 146–158). Hershey, PA: IGI Global. doi:10.4018/978-1-4666-8828-5.ch007

Doyle, D. J., & Fahy, P. J. (2018). Interactivity in Distance Education and Computer-Aided Learning, With Medical Education Examples. In M. Khosrow-Pour, D.B.A. (Ed.), Encyclopedia of Information Science and Technology, Fourth Edition (pp. 5829-5840). Hershey, PA: IGI Global. doi:10.4018/978-1-5225-2255-3.ch507

Elias, N. I., & Walker, T. W. (2017). Factors that Contribute to Continued Use of E-Training among Healthcare Professionals. In F. Topor (Ed.), *Handbook of Research on Individualism and Identity in the Globalized Digital Age* (pp. 403–429). Hershey, PA: IGI Global. doi:10.4018/978-1-5225-0522-8.ch018

Eloy, S., Dias, M. S., Lopes, P. F., & Vilar, E. (2016). Digital Technologies in Architecture and Engineering: Exploring an Engaged Interaction within Curricula. In D. Fonseca & E. Redondo (Eds.), *Handbook of Research on Applied E-Learning in Engineering and Architecture Education* (pp. 368–402). Hershey, PA: IGI Global. doi:10.4018/978-1-4666-8803-2.ch017

Estrela, V. V., Magalhães, H. A., & Saotome, O. (2016). Total Variation Applications in Computer Vision. In N. Kamila (Ed.), *Handbook of Research on Emerging Perspectives in Intelligent Pattern Recognition, Analysis, and Image Processing* (pp. 41–64). Hershey, PA: IGI Global. doi:10.4018/978-1-4666-8654-0.ch002

Filipovic, N., Radovic, M., Nikolic, D. D., Saveljic, I., Milosevic, Z., Exarchos, T. P., ... Parodi, O. (2016). Computer Predictive Model for Plaque Formation and Progression in the Artery. In D. Fotiadis (Ed.), *Handbook of Research on Trends in the Diagnosis and Treatment of Chronic Conditions* (pp. 279–300). Hershey, PA: IGI Global. doi:10.4018/978-1-4666-8828-5.ch013

Fisher, R. L. (2018). Computer-Assisted Indian Matrimonial Services. In M. Khosrow-Pour, D.B.A. (Ed.), Encyclopedia of Information Science and Technology, Fourth Edition (pp. 4136-4145). Hershey, PA: IGI Global. doi:10.4018/978-1-5225-2255-3.ch358

Fleenor, H. G., & Hodhod, R. (2016). Assessment of Learning and Technology: Computer Science Education. In V. Wang (Ed.), *Handbook of Research on Learning Outcomes and Opportunities in the Digital Age* (pp. 51–78). Hershey, PA: IGI Global. doi:10.4018/978-1-4666-9577-1.ch003

García-Valcárcel, A., & Mena, J. (2016). Information Technology as a Way To Support Collaborative Learning: What In-Service Teachers Think, Know and Do. *Journal of Information Technology Research*, 9(1), 1–17. doi:10.4018/JITR.2016010101

Gardner-McCune, C., & Jimenez, Y. (2017). Historical App Developers: Integrating CS into K-12 through Cross-Disciplinary Projects. In Y. Rankin & J. Thomas (Eds.), *Moving Students of Color from Consumers to Producers of Technology* (pp. 85–112). Hershey, PA: IGI Global. doi:10.4018/978-1-5225-2005-4.ch005

Garvey, G. P. (2016). Exploring Perception, Cognition, and Neural Pathways of Stereo Vision and the Split–Brain Human Computer Interface. In A. Ursyn (Ed.), *Knowledge Visualization and Visual Literacy in Science Education* (pp. 28–76). Hershey, PA: IGI Global. doi:10.4018/978-1-5225-0480-1.ch002

Ghafele, R., & Gibert, B. (2018). Open Growth: The Economic Impact of Open Source Software in the USA. In M. Khosrow-Pour (Ed.), *Optimizing Contemporary Application and Processes in Open Source Software* (pp. 164–197). Hershey, PA: IGI Global. doi:10.4018/978-1-5225-5314-4.ch007

Ghobakhloo, M., & Azar, A. (2018). Information Technology Resources, the Organizational Capability of Lean-Agile Manufacturing, and Business Performance. *Information Resources Management Journal*, *31*(2), 47–74. doi:10.4018/IRMJ.2018040103

Gianni, M., & Gotzamani, K. (2016). Integrated Management Systems and Information Management Systems: Common Threads. In P. Papajorgji, F. Pinet, A. Guimarães, & J. Papathanasiou (Eds.), *Automated Enterprise Systems for Maximizing Business Performance* (pp. 195–214). Hershey, PA: IGI Global. doi:10.4018/978-1-4666-8841-4.ch011

Gikandi, J. W. (2017). Computer-Supported Collaborative Learning and Assessment: A Strategy for Developing Online Learning Communities in Continuing Education. In J. Keengwe & G. Onchwari (Eds.), *Handbook of Research on Learner-Centered Pedagogy in Teacher Education and Professional Development* (pp. 309–333). Hershey, PA: IGI Global. doi:10.4018/978-1-5225-0892-2.ch017

Gokhale, A. A., & Machina, K. F. (2017). Development of a Scale to Measure Attitudes toward Information Technology. In L. Tomei (Ed.), *Exploring the New Era of Technology-Infused Education* (pp. 49–64). Hershey, PA: IGI Global. doi:10.4018/978-1-5225-1709-2.ch004

Grace, A., O'Donoghue, J., Mahony, C., Heffernan, T., Molony, D., & Carroll, T. (2016). Computerized Decision Support Systems for Multimorbidity Care: An Urgent Call for Research and Development. In M. Cruz-Cunha, I. Miranda, R. Martinho, & R. Rijo (Eds.), *Encyclopedia of E-Health and Telemedicine* (pp. 486–494). Hershey, PA: IGI Global. doi:10.4018/978-1-4666-9978-6.ch038

Gupta, A., & Singh, O. (2016). Computer Aided Modeling and Finite Element Analysis of Human Elbow. *International Journal of Biomedical and Clinical Engineering*, *5*(1), 31–38. doi:10.4018/IJBCE.2016010104

H., S. K. (2016). Classification of Cybercrimes and Punishments under the Information Technology Act, 2000. In S. Geetha, & A. Phamila (Eds.), *Combating Security Breaches and Criminal Activity in the Digital Sphere* (pp. 57-66). Hershey, PA: IGI Global. doi:10.4018/978-1-5225-0193-0.ch004

Hafeez-Baig, A., Gururajan, R., & Wickramasinghe, N. (2017). Readiness as a Novel Construct of Readiness Acceptance Model (RAM) for the Wireless Handheld Technology. In N. Wickramasinghe (Ed.), *Handbook of Research on Healthcare Administration and Management* (pp. 578–595). Hershey, PA: IGI Global. doi:10.4018/978-1-5225-0920-2.ch035

Hanafizadeh, P., Ghandchi, S., & Asgarimehr, M. (2017). Impact of Information Technology on Lifestyle: A Literature Review and Classification. *International Journal of Virtual Communities and Social Networking*, *9*(2), 1–23. doi:10.4018/IJVCSN.2017040101

Harlow, D. B., Dwyer, H., Hansen, A. K., Hill, C., Iveland, A., Leak, A. E., & Franklin, D. M. (2016). Computer Programming in Elementary and Middle School: Connections across Content. In M. Urban & D. Falvo (Eds.), *Improving K-12 STEM Education Outcomes through Technological Integration* (pp. 337–361). Hershey, PA: IGI Global. doi:10.4018/978-1-4666-9616-7.ch015

Haseski, H. İ., Ilic, U., & Tuğtekin, U. (2018). Computational Thinking in Educational Digital Games: An Assessment Tool Proposal. In H. Ozcinar, G. Wong, & H. Ozturk (Eds.), *Teaching Computational Thinking in Primary Education* (pp. 256–287). Hershey, PA: IGI Global. doi:10.4018/978-1-5225-3200-2.ch013

Hee, W. J., Jalleh, G., Lai, H., & Lin, C. (2017). E-Commerce and IT Projects: Evaluation and Management Issues in Australian and Taiwanese Hospitals. *International Journal of Public Health Management and Ethics*, 2(1), 69–90. doi:10.4018/IJPHME.2017010104

Hernandez, A. A. (2017). Green Information Technology Usage: Awareness and Practices of Philippine IT Professionals. *International Journal of Enterprise Information Systems*, 13(4), 90–103. doi:10.4018/IJEIS.2017100106

Hernandez, A. A., & Ona, S. E. (2016). Green IT Adoption: Lessons from the Philippines Business Process Outsourcing Industry. *International Journal of Social Ecology and Sustainable Development*, 7(1), 1–34. doi:10.4018/IJSESD.2016010101

Hernandez, M. A., Marin, E. C., Garcia-Rodriguez, J., Azorin-Lopez, J., & Cazorla, M. (2017). Automatic Learning Improves Human-Robot Interaction in Productive Environments: A Review. *International Journal of Computer Vision and Image Processing*, 7(3), 65–75. doi:10.4018/IJCVIP.2017070106

Horne-Popp, L. M., Tessone, E. B., & Welker, J. (2018). If You Build It, They Will Come: Creating a Library Statistics Dashboard for Decision-Making. In L. Costello & M. Powers (Eds.), *Developing In-House Digital Tools in Library Spaces* (pp. 177–203). Hershey, PA: IGI Global. doi:10.4018/978-1-5225-2676-6.ch009

Hossan, C. G., & Ryan, J. C. (2016). Factors Affecting e-Government Technology Adoption Behaviour in a Voluntary Environment. *International Journal of Electronic Government Research*, 12(1), 24–49. doi:10.4018/IJEGR.2016010102

Hu, H., Hu, P. J., & Al-Gahtani, S. S. (2017). User Acceptance of Computer Technology at Work in Arabian Culture: A Model Comparison Approach. In M. Khosrow-Pour (Ed.), *Handbook of Research on Technology Adoption, Social Policy, and Global Integration* (pp. 205–228). Hershey, PA: IGI Global. doi:10.4018/978-1-5225-2668-1.ch011

Huie, C. P. (2016). Perceptions of Business Intelligence Professionals about Factors Related to Business Intelligence input in Decision Making. *International Journal of Business Analytics*, 3(3), 1–24. doi:10.4018/IJBAN.2016070101

Hung, S., Huang, W., Yen, D. C., Chang, S., & Lu, C. (2016). Effect of Information Service Competence and Contextual Factors on the Effectiveness of Strategic Information Systems Planning in Hospitals. *Journal of Global Information Management*, 24(1), 14–36. doi:10.4018/JGIM.2016010102

Ifinedo, P. (2017). Using an Extended Theory of Planned Behavior to Study Nurses' Adoption of Healthcare Information Systems in Nova Scotia. *International Journal of Technology Diffusion*, 8(1), 1–17. doi:10.4018/IJTD.2017010101

Ilie, V., & Sneha, S. (2018). A Three Country Study for Understanding Physicians' Engagement With Electronic Information Resources Pre and Post System Implementation. *Journal of Global Information Management*, 26(2), 48–73. doi:10.4018/JGIM.2018040103

Inoue-Smith, Y. (2017). Perceived Ease in Using Technology Predicts Teacher Candidates' Preferences for Online Resources. *International Journal of Online Pedagogy and Course Design*, 7(3), 17–28. doi:10.4018/IJOPCD.2017070102

Islam, A. A. (2016). Development and Validation of the Technology Adoption and Gratification (TAG) Model in Higher Education: A Cross-Cultural Study Between Malaysia and China. *International Journal of Technology and Human Interaction, 12*(3), 78–105. doi:10.4018/IJTHI.2016070106

Islam, A. Y. (2017). Technology Satisfaction in an Academic Context: Moderating Effect of Gender. In A. Mesquita (Ed.), *Research Paradigms and Contemporary Perspectives on Human-Technology Interaction* (pp. 187–211). Hershey, PA: IGI Global. doi:10.4018/978-1-5225-1868-6.ch009

Jamil, G. L., & Jamil, C. C. (2017). Information and Knowledge Management Perspective Contributions for Fashion Studies: Observing Logistics and Supply Chain Management Processes. In G. Jamil, A. Soares, & C. Pessoa (Eds.), *Handbook of Research on Information Management for Effective Logistics and Supply Chains* (pp. 199–221). Hershey, PA: IGI Global. doi:10.4018/978-1-5225-0973-8.ch011

Jamil, G. L., Jamil, L. C., Vieira, A. A., & Xavier, A. J. (2016). Challenges in Modelling Healthcare Services: A Study Case of Information Architecture Perspectives. In G. Jamil, J. Poças Rascão, F. Ribeiro, & A. Malheiro da Silva (Eds.), *Handbook of Research on Information Architecture and Management in Modern Organizations* (pp. 1–23). Hershey, PA: IGI Global. doi:10.4018/978-1-4666-8637-3.ch001

Janakova, M. (2018). Big Data and Simulations for the Solution of Controversies in Small Businesses. In M. Khosrow-Pour, D.B.A. (Ed.), Encyclopedia of Information Science and Technology, Fourth Edition (pp. 6907-6915). Hershey, PA: IGI Global. doi:10.4018/978-1-5225-2255-3.ch598

Jha, D. G. (2016). Preparing for Information Technology Driven Changes. In S. Tiwari & L. Nafees (Eds.), *Innovative Management Education Pedagogies for Preparing Next-Generation Leaders* (pp. 258–274). Hershey, PA: IGI Global. doi:10.4018/978-1-4666-9691-4.ch015

Jhawar, A., & Garg, S. K. (2018). Logistics Improvement by Investment in Information Technology Using System Dynamics. In A. Azar & S. Vaidyanathan (Eds.), *Advances in System Dynamics and Control* (pp. 528–567). Hershey, PA: IGI Global. doi:10.4018/978-1-5225-4077-9.ch017

Kalelioğlu, F., Gülbahar, Y., & Doğan, D. (2018). Teaching How to Think Like a Programmer: Emerging Insights. In H. Ozcinar, G. Wong, & H. Ozturk (Eds.), *Teaching Computational Thinking in Primary Education* (pp. 18–35). Hershey, PA: IGI Global. doi:10.4018/978-1-5225-3200-2.ch002

Kamberi, S. (2017). A Girls-Only Online Virtual World Environment and its Implications for Game-Based Learning. In A. Stricker, C. Calongne, B. Truman, & F. Arenas (Eds.), *Integrating an Awareness of Selfhood and Society into Virtual Learning* (pp. 74–95). Hershey, PA: IGI Global. doi:10.4018/978-1-5225-2182-2.ch006

Kamel, S., & Rizk, N. (2017). ICT Strategy Development: From Design to Implementation – Case of Egypt. In C. Howard & K. Hargiss (Eds.), *Strategic Information Systems and Technologies in Modern Organizations* (pp. 239–257). Hershey, PA: IGI Global. doi:10.4018/978-1-5225-1680-4.ch010

Kamel, S. H. (2018). The Potential Role of the Software Industry in Supporting Economic Development. In M. Khosrow-Pour, D.B.A. (Ed.), Encyclopedia of Information Science and Technology, Fourth Edition (pp. 7259-7269). Hershey, PA: IGI Global. doi:10.4018/978-1-5225-2255-3.ch631

Karon, R. (2016). Utilisation of Health Information Systems for Service Delivery in the Namibian Environment. In T. Iyamu & A. Tatnall (Eds.), *Maximizing Healthcare Delivery and Management through Technology Integration* (pp. 169–183). Hershey, PA: IGI Global. doi:10.4018/978-1-4666-9446-0.ch011

Kawata, S. (2018). Computer-Assisted Parallel Program Generation. In M. Khosrow-Pour, D.B.A. (Ed.), Encyclopedia of Information Science and Technology, Fourth Edition (pp. 4583-4593). Hershey, PA: IGI Global. doi:10.4018/978-1-5225-2255-3.ch398

Khanam, S., Siddiqui, J., & Talib, F. (2016). A DEMATEL Approach for Prioritizing the TQM Enablers and IT Resources in the Indian ICT Industry. *International Journal of Applied Management Sciences and Engineering*, *3*(1), 11–29. doi:10.4018/IJAMSE.2016010102

Khari, M., Shrivastava, G., Gupta, S., & Gupta, R. (2017). Role of Cyber Security in Today's Scenario. In R. Kumar, P. Pattnaik, & P. Pandey (Eds.), *Detecting and Mitigating Robotic Cyber Security Risks* (pp. 177–191). Hershey, PA: IGI Global. doi:10.4018/978-1-5225-2154-9.ch013

Khouja, M., Rodriguez, I. B., Ben Halima, Y., & Moalla, S. (2018). IT Governance in Higher Education Institutions: A Systematic Literature Review. *International Journal of Human Capital and Information Technology Professionals*, *9*(2), 52–67. doi:10.4018/IJHCITP.2018040104

Kim, S., Chang, M., Choi, N., Park, J., & Kim, H. (2016). The Direct and Indirect Effects of Computer Uses on Student Success in Math. *International Journal of Cyber Behavior, Psychology and Learning*, *6*(3), 48–64. doi:10.4018/IJCBPL.2016070104

Kiourt, C., Pavlidis, G., Koutsoudis, A., & Kalles, D. (2017). Realistic Simulation of Cultural Heritage. *International Journal of Computational Methods in Heritage Science*, *1*(1), 10–40. doi:10.4018/IJCMHS.2017010102

Korikov, A., & Krivtsov, O. (2016). System of People-Computer: On the Way of Creation of Human-Oriented Interface. In V. Mkrttchian, A. Bershadsky, A. Bozhday, M. Kataev, & S. Kataev (Eds.), *Handbook of Research on Estimation and Control Techniques in E-Learning Systems* (pp. 458–470). Hershey, PA: IGI Global. doi:10.4018/978-1-4666-9489-7.ch032

Köse, U. (2017). An Augmented-Reality-Based Intelligent Mobile Application for Open Computer Education. In G. Kurubacak & H. Altinpulluk (Eds.), *Mobile Technologies and Augmented Reality in Open Education* (pp. 154–174). Hershey, PA: IGI Global. doi:10.4018/978-1-5225-2110-5.ch008

Lahmiri, S. (2018). Information Technology Outsourcing Risk Factors and Provider Selection. In M. Gupta, R. Sharman, J. Walp, & P. Mulgund (Eds.), *Information Technology Risk Management and Compliance in Modern Organizations* (pp. 214–228). Hershey, PA: IGI Global. doi:10.4018/978-1-5225-2604-9.ch008

Landriscina, F. (2017). Computer-Supported Imagination: The Interplay Between Computer and Mental Simulation in Understanding Scientific Concepts. In I. Levin & D. Tsybulsky (Eds.), *Digital Tools and Solutions for Inquiry-Based STEM Learning* (pp. 33–60). Hershey, PA: IGI Global. doi:10.4018/978-1-5225-2525-7.ch002

Lau, S. K., Winley, G. K., Leung, N. K., Tsang, N., & Lau, S. Y. (2016). An Exploratory Study of Expectation in IT Skills in a Developing Nation: Vietnam. *Journal of Global Information Management*, *24*(1), 1–13. doi:10.4018/JGIM.2016010101

Lavranos, C., Kostagiolas, P., & Papadatos, J. (2016). Information Retrieval Technologies and the "Realities" of Music Information Seeking. In I. Deliyannis, P. Kostagiolas, & C. Banou (Eds.), *Experimental Multimedia Systems for Interactivity and Strategic Innovation* (pp. 102–121). Hershey, PA: IGI Global. doi:10.4018/978-1-4666-8659-5.ch005

Lee, W. W. (2018). Ethical Computing Continues From Problem to Solution. In M. Khosrow-Pour, D.B.A. (Ed.), Encyclopedia of Information Science and Technology, Fourth Edition (pp. 4884-4897). Hershey, PA: IGI Global. doi:10.4018/978-1-5225-2255-3.ch423

Lehto, M. (2016). Cyber Security Education and Research in the Finland's Universities and Universities of Applied Sciences. *International Journal of Cyber Warfare & Terrorism*, *6*(2), 15–31. doi:10.4018/IJCWT.2016040102

Lin, C., Jalleh, G., & Huang, Y. (2016). Evaluating and Managing Electronic Commerce and Outsourcing Projects in Hospitals. In A. Dwivedi (Ed.), *Reshaping Medical Practice and Care with Health Information Systems* (pp. 132–172). Hershey, PA: IGI Global. doi:10.4018/978-1-4666-9870-3.ch005

Lin, S., Chen, S., & Chuang, S. (2017). Perceived Innovation and Quick Response Codes in an Online-to-Offline E-Commerce Service Model. *International Journal of E-Adoption*, *9*(2), 1–16. doi:10.4018/IJEA.2017070101

Liu, M., Wang, Y., Xu, W., & Liu, L. (2017). Automated Scoring of Chinese Engineering Students' English Essays. *International Journal of Distance Education Technologies*, *15*(1), 52–68. doi:10.4018/IJDET.2017010104

Luciano, E. M., Wiedenhöft, G. C., Macadar, M. A., & Pinheiro dos Santos, F. (2016). Information Technology Governance Adoption: Understanding its Expectations Through the Lens of Organizational Citizenship. *International Journal of IT/Business Alignment and Governance, 7*(2), 22-32. doi:10.4018/IJITBAG.2016070102

Mabe, L. K., & Oladele, O. I. (2017). Application of Information Communication Technologies for Agricultural Development through Extension Services: A Review. In T. Tossy (Ed.), *Information Technology Integration for Socio-Economic Development* (pp. 52–101). Hershey, PA: IGI Global. doi:10.4018/978-1-5225-0539-6.ch003

Manogaran, G., Thota, C., & Lopez, D. (2018). Human-Computer Interaction With Big Data Analytics. In D. Lopez & M. Durai (Eds.), *HCI Challenges and Privacy Preservation in Big Data Security* (pp. 1–22). Hershey, PA: IGI Global. doi:10.4018/978-1-5225-2863-0.ch001

Margolis, J., Goode, J., & Flapan, J. (2017). A Critical Crossroads for Computer Science for All: "Identifying Talent" or "Building Talent," and What Difference Does It Make? In Y. Rankin & J. Thomas (Eds.), *Moving Students of Color from Consumers to Producers of Technology* (pp. 1–23). Hershey, PA: IGI Global. doi:10.4018/978-1-5225-2005-4.ch001

Mbale, J. (2018). Computer Centres Resource Cloud Elasticity-Scalability (CRECES): Copperbelt University Case Study. In S. Aljawarneh & M. Malhotra (Eds.), *Critical Research on Scalability and Security Issues in Virtual Cloud Environments* (pp. 48–70). Hershey, PA: IGI Global. doi:10.4018/978-1-5225-3029-9.ch003

McKee, J. (2018). The Right Information: The Key to Effective Business Planning. In *Business Architectures for Risk Assessment and Strategic Planning: Emerging Research and Opportunities* (pp. 38–52). Hershey, PA: IGI Global. doi:10.4018/978-1-5225-3392-4.ch003

Mensah, I. K., & Mi, J. (2018). Determinants of Intention to Use Local E-Government Services in Ghana: The Perspective of Local Government Workers. *International Journal of Technology Diffusion, 9*(2), 41–60. doi:10.4018/IJTD.2018040103

Mohamed, J. H. (2018). Scientograph-Based Visualization of Computer Forensics Research Literature. In J. Jeyasekar & P. Saravanan (Eds.), *Innovations in Measuring and Evaluating Scientific Information* (pp. 148–162). Hershey, PA: IGI Global. doi:10.4018/978-1-5225-3457-0.ch010

Moore, R. L., & Johnson, N. (2017). Earning a Seat at the Table: How IT Departments Can Partner in Organizational Change and Innovation. *International Journal of Knowledge-Based Organizations, 7*(2), 1–12. doi:10.4018/IJKBO.2017040101

Mtebe, J. S., & Kissaka, M. M. (2016). Enhancing the Quality of Computer Science Education with MOOCs in Sub-Saharan Africa. In J. Keengwe & G. Onchwari (Eds.), *Handbook of Research on Active Learning and the Flipped Classroom Model in the Digital Age* (pp. 366–377). Hershey, PA: IGI Global. doi:10.4018/978-1-4666-9680-8.ch019

Mukul, M. K., & Bhattaharyya, S. (2017). Brain-Machine Interface: Human-Computer Interaction. In E. Noughabi, B. Raahemi, A. Albadvi, & B. Far (Eds.), *Handbook of Research on Data Science for Effective Healthcare Practice and Administration* (pp. 417–443). Hershey, PA: IGI Global. doi:10.4018/978-1-5225-2515-8.ch018

Na, L. (2017). Library and Information Science Education and Graduate Programs in Academic Libraries. In L. Ruan, Q. Zhu, & Y. Ye (Eds.), *Academic Library Development and Administration in China* (pp. 218–229). Hershey, PA: IGI Global. doi:10.4018/978-1-5225-0550-1.ch013

Nabavi, A., Taghavi-Fard, M. T., Hanafizadeh, P., & Taghva, M. R. (2016). Information Technology Continuance Intention: A Systematic Literature Review. *International Journal of E-Business Research, 12*(1), 58–95. doi:10.4018/IJEBR.2016010104

Nath, R., & Murthy, V. N. (2018). What Accounts for the Differences in Internet Diffusion Rates Around the World? In M. Khosrow-Pour, D.B.A. (Ed.), Encyclopedia of Information Science and Technology, Fourth Edition (pp. 8095-8104). Hershey, PA: IGI Global. doi:10.4018/978-1-5225-2255-3.ch705

Nedelko, Z., & Potocan, V. (2018). The Role of Emerging Information Technologies for Supporting Supply Chain Management. In M. Khosrow-Pour, D.B.A. (Ed.), Encyclopedia of Information Science and Technology, Fourth Edition (pp. 5559-5569). Hershey, PA: IGI Global. doi:10.4018/978-1-5225-2255-3.ch483

Ngafeeson, M. N. (2018). User Resistance to Health Information Technology. In M. Khosrow-Pour, D.B.A. (Ed.), Encyclopedia of Information Science and Technology, Fourth Edition (pp. 3816-3825). Hershey, PA: IGI Global. doi:10.4018/978-1-5225-2255-3.ch331

Nozari, H., Najafi, S. E., Jafari-Eskandari, M., & Aliahmadi, A. (2016). Providing a Model for Virtual Project Management with an Emphasis on IT Projects. In C. Graham (Ed.), *Strategic Management and Leadership for Systems Development in Virtual Spaces* (pp. 43–63). Hershey, PA: IGI Global. doi:10.4018/978-1-4666-9688-4.ch003

Nurdin, N., Stockdale, R., & Scheepers, H. (2016). Influence of Organizational Factors in the Sustainability of E-Government: A Case Study of Local E-Government in Indonesia. In I. Sodhi (Ed.), *Trends, Prospects, and Challenges in Asian E-Governance* (pp. 281–323). Hershey, PA: IGI Global. doi:10.4018/978-1-4666-9536-8.ch014

Odagiri, K. (2017). Introduction of Individual Technology to Constitute the Current Internet. In *Strategic Policy-Based Network Management in Contemporary Organizations* (pp. 20–96). Hershey, PA: IGI Global. doi:10.4018/978-1-68318-003-6.ch003

Okike, E. U. (2018). Computer Science and Prison Education. In I. Biao (Ed.), *Strategic Learning Ideologies in Prison Education Programs* (pp. 246–264). Hershey, PA: IGI Global. doi:10.4018/978-1-5225-2909-5.ch012

Olelewe, C. J., & Nwafor, I. P. (2017). Level of Computer Appreciation Skills Acquired for Sustainable Development by Secondary School Students in Nsukka LGA of Enugu State, Nigeria. In C. Ayo & V. Mbarika (Eds.), *Sustainable ICT Adoption and Integration for Socio-Economic Development* (pp. 214–233). Hershey, PA: IGI Global. doi:10.4018/978-1-5225-2565-3.ch010

Oliveira, M., Maçada, A. C., Curado, C., & Nodari, F. (2017). Infrastructure Profiles and Knowledge Sharing. *International Journal of Technology and Human Interaction*, *13*(3), 1–12. doi:10.4018/IJTHI.2017070101

Otarkhani, A., Shokouhyar, S., & Pour, S. S. (2017). Analyzing the Impact of Governance of Enterprise IT on Hospital Performance: Tehran's (Iran) Hospitals – A Case Study. *International Journal of Healthcare Information Systems and Informatics*, *12*(3), 1–20. doi:10.4018/IJHISI.2017070101

Otunla, A. O., & Amuda, C. O. (2018). Nigerian Undergraduate Students' Computer Competencies and Use of Information Technology Tools and Resources for Study Skills and Habits' Enhancement. In M. Khosrow-Pour, D.B.A. (Ed.), Encyclopedia of Information Science and Technology, Fourth Edition (pp. 2303-2313). Hershey, PA: IGI Global. doi:10.4018/978-1-5225-2255-3.ch200

Özçınar, H. (2018). A Brief Discussion on Incentives and Barriers to Computational Thinking Education. In H. Ozcinar, G. Wong, & H. Ozturk (Eds.), *Teaching Computational Thinking in Primary Education* (pp. 1–17). Hershey, PA: IGI Global. doi:10.4018/978-1-5225-3200-2.ch001

Pandey, J. M., Garg, S., Mishra, P., & Mishra, B. P. (2017). Computer Based Psychological Interventions: Subject to the Efficacy of Psychological Services. *International Journal of Computers in Clinical Practice*, *2*(1), 25–33. doi:10.4018/IJCCP.2017010102

Parry, V. K., & Lind, M. L. (2016). Alignment of Business Strategy and Information Technology Considering Information Technology Governance, Project Portfolio Control, and Risk Management. *International Journal of Information Technology Project Management*, *7*(4), 21–37. doi:10.4018/IJITPM.2016100102

Patro, C. (2017). Impulsion of Information Technology on Human Resource Practices. In P. Ordóñez de Pablos (Ed.), *Managerial Strategies and Solutions for Business Success in Asia* (pp. 231–254). Hershey, PA: IGI Global. doi:10.4018/978-1-5225-1886-0.ch013

Patro, C. S., & Raghunath, K. M. (2017). Information Technology Paraphernalia for Supply Chain Management Decisions. In M. Tavana (Ed.), *Enterprise Information Systems and the Digitalization of Business Functions* (pp. 294–320). Hershey, PA: IGI Global. doi:10.4018/978-1-5225-2382-6.ch014

Paul, P. K. (2016). Cloud Computing: An Agent of Promoting Interdisciplinary Sciences, Especially Information Science and I-Schools – Emerging Techno-Educational Scenario. In L. Chao (Ed.), *Handbook of Research on Cloud-Based STEM Education for Improved Learning Outcomes* (pp. 247–258). Hershey, PA: IGI Global. doi:10.4018/978-1-4666-9924-3.ch016

Paul, P. K. (2018). The Context of IST for Solid Information Retrieval and Infrastructure Building: Study of Developing Country. *International Journal of Information Retrieval Research*, 8(1), 86–100. doi:10.4018/IJIRR.2018010106

Paul, P. K., & Chatterjee, D. (2018). iSchools Promoting "Information Science and Technology" (IST) Domain Towards Community, Business, and Society With Contemporary Worldwide Trend and Emerging Potentialities in India. In M. Khosrow-Pour, D.B.A. (Ed.), Encyclopedia of Information Science and Technology, Fourth Edition (pp. 4723-4735). Hershey, PA: IGI Global. doi:10.4018/978-1-5225-2255-3.ch410

Pessoa, C. R., & Marques, M. E. (2017). Information Technology and Communication Management in Supply Chain Management. In G. Jamil, A. Soares, & C. Pessoa (Eds.), *Handbook of Research on Information Management for Effective Logistics and Supply Chains* (pp. 23–33). Hershey, PA: IGI Global. doi:10.4018/978-1-5225-0973-8.ch002

Pineda, R. G. (2016). Where the Interaction Is Not: Reflections on the Philosophy of Human-Computer Interaction. *International Journal of Art, Culture and Design Technologies*, 5(1), 1–12. doi:10.4018/IJACDT.2016010101

Pineda, R. G. (2018). Remediating Interaction: Towards a Philosophy of Human-Computer Relationship. In M. Khosrow-Pour (Ed.), *Enhancing Art, Culture, and Design With Technological Integration* (pp. 75–98). Hershey, PA: IGI Global. doi:10.4018/978-1-5225-5023-5.ch004

Poikela, P., & Vuojärvi, H. (2016). Learning ICT-Mediated Communication through Computer-Based Simulations. In M. Cruz-Cunha, I. Miranda, R. Martinho, & R. Rijo (Eds.), *Encyclopedia of E-Health and Telemedicine* (pp. 674–687). Hershey, PA: IGI Global. doi:10.4018/978-1-4666-9978-6.ch052

Qian, Y. (2017). Computer Simulation in Higher Education: Affordances, Opportunities, and Outcomes. In P. Vu, S. Fredrickson, & C. Moore (Eds.), *Handbook of Research on Innovative Pedagogies and Technologies for Online Learning in Higher Education* (pp. 236–262). Hershey, PA: IGI Global. doi:10.4018/978-1-5225-1851-8.ch011

Radant, O., Colomo-Palacios, R., & Stantchev, V. (2016). Factors for the Management of Scarce Human Resources and Highly Skilled Employees in IT-Departments: A Systematic Review. *Journal of Information Technology Research*, 9(1), 65–82. doi:10.4018/JITR.2016010105

Rahman, N. (2016). Toward Achieving Environmental Sustainability in the Computer Industry. *International Journal of Green Computing*, 7(1), 37–54. doi:10.4018/IJGC.2016010103

Rahman, N. (2017). Lessons from a Successful Data Warehousing Project Management. *International Journal of Information Technology Project Management*, 8(4), 30–45. doi:10.4018/IJITPM.2017100103

Rahman, N. (2018). Environmental Sustainability in the Computer Industry for Competitive Advantage. In M. Khosrow-Pour (Ed.), *Green Computing Strategies for Competitive Advantage and Business Sustainability* (pp. 110–130). Hershey, PA: IGI Global. doi:10.4018/978-1-5225-5017-4.ch006

Rajh, A., & Pavetic, T. (2017). Computer Generated Description as the Required Digital Competence in Archival Profession. *International Journal of Digital Literacy and Digital Competence*, 8(1), 36–49. doi:10.4018/IJDLDC.2017010103

Raman, A., & Goyal, D. P. (2017). Extending IMPLEMENT Framework for Enterprise Information Systems Implementation to Information System Innovation. In M. Tavana (Ed.), *Enterprise Information Systems and the Digitalization of Business Functions* (pp. 137–177). Hershey, PA: IGI Global. doi:10.4018/978-1-5225-2382-6.ch007

Rao, Y. S., Rauta, A. K., Saini, H., & Panda, T. C. (2017). Mathematical Model for Cyber Attack in Computer Network. *International Journal of Business Data Communications and Networking*, 13(1), 58–65. doi:10.4018/IJBDCN.2017010105

Rapaport, W. J. (2018). Syntactic Semantics and the Proper Treatment of Computationalism. In M. Danesi (Ed.), *Empirical Research on Semiotics and Visual Rhetoric* (pp. 128–176). Hershey, PA: IGI Global. doi:10.4018/978-1-5225-5622-0.ch007

Raut, R., Priyadarshinee, P., & Jha, M. (2017). Understanding the Mediation Effect of Cloud Computing Adoption in Indian Organization: Integrating TAM-TOE- Risk Model. *International Journal of Service Science, Management, Engineering, and Technology*, 8(3), 40–59. doi:10.4018/IJSSMET.2017070103

Regan, E. A., & Wang, J. (2016). Realizing the Value of EHR Systems Critical Success Factors. *International Journal of Healthcare Information Systems and Informatics*, 11(3), 1–18. doi:10.4018/IJHISI.2016070101

Rezaie, S., Mirabedini, S. J., & Abtahi, A. (2018). Designing a Model for Implementation of Business Intelligence in the Banking Industry. *International Journal of Enterprise Information Systems*, 14(1), 77–103. doi:10.4018/IJEIS.2018010105

Rezende, D. A. (2016). Digital City Projects: Information and Public Services Offered by Chicago (USA) and Curitiba (Brazil). *International Journal of Knowledge Society Research*, 7(3), 16–30. doi:10.4018/IJKSR.2016070102

Rezende, D. A. (2018). Strategic Digital City Projects: Innovative Information and Public Services Offered by Chicago (USA) and Curitiba (Brazil). In M. Lytras, L. Daniela, & A. Visvizi (Eds.), *Enhancing Knowledge Discovery and Innovation in the Digital Era* (pp. 204–223). Hershey, PA: IGI Global. doi:10.4018/978-1-5225-4191-2.ch012

Riabov, V. V. (2016). Teaching Online Computer-Science Courses in LMS and Cloud Environment. *International Journal of Quality Assurance in Engineering and Technology Education*, 5(4), 12–41. doi:10.4018/IJQAETE.2016100102

Ricordel, V., Wang, J., Da Silva, M. P., & Le Callet, P. (2016). 2D and 3D Visual Attention for Computer Vision: Concepts, Measurement, and Modeling. In R. Pal (Ed.), *Innovative Research in Attention Modeling and Computer Vision Applications* (pp. 1–44). Hershey, PA: IGI Global. doi:10.4018/978-1-4666-8723-3.ch001

Rodriguez, A., Rico-Diaz, A. J., Rabuñal, J. R., & Gestal, M. (2017). Fish Tracking with Computer Vision Techniques: An Application to Vertical Slot Fishways. In M. S., & V. V. (Eds.), Multi-Core Computer Vision and Image Processing for Intelligent Applications (pp. 74-104). Hershey, PA: IGI Global. doi:10.4018/978-1-5225-0889-2.ch003

Romero, J. A. (2018). Sustainable Advantages of Business Value of Information Technology. In M. Khosrow-Pour, D.B.A. (Ed.), Encyclopedia of Information Science and Technology, Fourth Edition (pp. 923-929). Hershey, PA: IGI Global. doi:10.4018/978-1-5225-2255-3.ch079

Romero, J. A. (2018). The Always-On Business Model and Competitive Advantage. In N. Bajgoric (Ed.), *Always-On Enterprise Information Systems for Modern Organizations* (pp. 23–40). Hershey, PA: IGI Global. doi:10.4018/978-1-5225-3704-5.ch002

Rosen, Y. (2018). Computer Agent Technologies in Collaborative Learning and Assessment. In M. Khosrow-Pour, D.B.A. (Ed.), Encyclopedia of Information Science and Technology, Fourth Edition (pp. 2402-2410). Hershey, PA: IGI Global. doi:10.4018/978-1-5225-2255-3.ch209

Rosen, Y., & Mosharraf, M. (2016). Computer Agent Technologies in Collaborative Assessments. In Y. Rosen, S. Ferrara, & M. Mosharraf (Eds.), *Handbook of Research on Technology Tools for Real-World Skill Development* (pp. 319–343). Hershey, PA: IGI Global. doi:10.4018/978-1-4666-9441-5.ch012

Roy, D. (2018). Success Factors of Adoption of Mobile Applications in Rural India: Effect of Service Characteristics on Conceptual Model. In M. Khosrow-Pour (Ed.), *Green Computing Strategies for Competitive Advantage and Business Sustainability* (pp. 211–238). Hershey, PA: IGI Global. doi:10.4018/978-1-5225-5017-4.ch010

Ruffin, T. R. (2016). Health Information Technology and Change. In V. Wang (Ed.), *Handbook of Research on Advancing Health Education through Technology* (pp. 259–285). Hershey, PA: IGI Global. doi:10.4018/978-1-4666-9494-1.ch012

Ruffin, T. R. (2016). Health Information Technology and Quality Management. *International Journal of Information Communication Technologies and Human Development*, 8(4), 56–72. doi:10.4018/IJI-CTHD.2016100105

Ruffin, T. R., & Hawkins, D. P. (2018). Trends in Health Care Information Technology and Informatics. In M. Khosrow-Pour, D.B.A. (Ed.), Encyclopedia of Information Science and Technology, Fourth Edition (pp. 3805-3815). Hershey, PA: IGI Global. doi:10.4018/978-1-5225-2255-3.ch330

Safari, M. R., & Jiang, Q. (2018). The Theory and Practice of IT Governance Maturity and Strategies Alignment: Evidence From Banking Industry. *Journal of Global Information Management, 26*(2), 127–146. doi:10.4018/JGIM.2018040106

Sahin, H. B., & Anagun, S. S. (2018). Educational Computer Games in Math Teaching: A Learning Culture. In E. Toprak & E. Kumtepe (Eds.), *Supporting Multiculturalism in Open and Distance Learning Spaces* (pp. 249–280). Hershey, PA: IGI Global. doi:10.4018/978-1-5225-3076-3.ch013

Sanna, A., & Valpreda, F. (2017). An Assessment of the Impact of a Collaborative Didactic Approach and Students' Background in Teaching Computer Animation. *International Journal of Information and Communication Technology Education, 13*(4), 1–16. doi:10.4018/IJICTE.2017100101

Savita, K., Dominic, P., & Ramayah, T. (2016). The Drivers, Practices and Outcomes of Green Supply Chain Management: Insights from ISO14001 Manufacturing Firms in Malaysia. *International Journal of Information Systems and Supply Chain Management, 9*(2), 35–60. doi:10.4018/IJISSCM.2016040103

Scott, A., Martin, A., & McAlear, F. (2017). Enhancing Participation in Computer Science among Girls of Color: An Examination of a Preparatory AP Computer Science Intervention. In Y. Rankin & J. Thomas (Eds.), *Moving Students of Color from Consumers to Producers of Technology* (pp. 62–84). Hershey, PA: IGI Global. doi:10.4018/978-1-5225-2005-4.ch004

Shahsavandi, E., Mayah, G., & Rahbari, H. (2016). Impact of E-Government on Transparency and Corruption in Iran. In I. Sodhi (Ed.), *Trends, Prospects, and Challenges in Asian E-Governance* (pp. 75–94). Hershey, PA: IGI Global. doi:10.4018/978-1-4666-9536-8.ch004

Siddoo, V., & Wongsai, N. (2017). Factors Influencing the Adoption of ISO/IEC 29110 in Thai Government Projects: A Case Study. *International Journal of Information Technologies and Systems Approach, 10*(1), 22–44. doi:10.4018/IJITSA.2017010102

Sidorkina, I., & Rybakov, A. (2016). Computer-Aided Design as Carrier of Set Development Changes System in E-Course Engineering. In V. Mkrttchian, A. Bershadsky, A. Bozhday, M. Kataev, & S. Kataev (Eds.), *Handbook of Research on Estimation and Control Techniques in E-Learning Systems* (pp. 500–515). Hershey, PA: IGI Global. doi:10.4018/978-1-4666-9489-7.ch035

Sidorkina, I., & Rybakov, A. (2016). Creating Model of E-Course: As an Object of Computer-Aided Design. In V. Mkrttchian, A. Bershadsky, A. Bozhday, M. Kataev, & S. Kataev (Eds.), *Handbook of Research on Estimation and Control Techniques in E-Learning Systems* (pp. 286–297). Hershey, PA: IGI Global. doi:10.4018/978-1-4666-9489-7.ch019

Simões, A. (2017). Using Game Frameworks to Teach Computer Programming. In R. Alexandre Peixoto de Queirós & M. Pinto (Eds.), *Gamification-Based E-Learning Strategies for Computer Programming Education* (pp. 221–236). Hershey, PA: IGI Global. doi:10.4018/978-1-5225-1034-5.ch010

Sllame, A. M. (2017). Integrating LAB Work With Classes in Computer Network Courses. In H. Alphin Jr, R. Chan, & J. Lavine (Eds.), *The Future of Accessibility in International Higher Education* (pp. 253–275). Hershey, PA: IGI Global. doi:10.4018/978-1-5225-2560-8.ch015

Smirnov, A., Ponomarev, A., Shilov, N., Kashevnik, A., & Teslya, N. (2018). Ontology-Based Human-Computer Cloud for Decision Support: Architecture and Applications in Tourism. *International Journal of Embedded and Real-Time Communication Systems*, *9*(1), 1–19. doi:10.4018/IJERTCS.2018010101

Smith-Ditizio, A. A., & Smith, A. D. (2018). Computer Fraud Challenges and Its Legal Implications. In M. Khosrow-Pour, D.B.A. (Ed.), Encyclopedia of Information Science and Technology, Fourth Edition (pp. 4837-4848). Hershey, PA: IGI Global. doi:10.4018/978-1-5225-2255-3.ch419

Sohani, S. S. (2016). Job Shadowing in Information Technology Projects: A Source of Competitive Advantage. *International Journal of Information Technology Project Management*, *7*(1), 47–57. doi:10.4018/IJITPM.2016010104

Sosnin, P. (2018). Figuratively Semantic Support of Human-Computer Interactions. In *Experience-Based Human-Computer Interactions: Emerging Research and Opportunities* (pp. 244–272). Hershey, PA: IGI Global. doi:10.4018/978-1-5225-2987-3.ch008

Spinelli, R., & Benevolo, C. (2016). From Healthcare Services to E-Health Applications: A Delivery System-Based Taxonomy. In A. Dwivedi (Ed.), *Reshaping Medical Practice and Care with Health Information Systems* (pp. 205–245). Hershey, PA: IGI Global. doi:10.4018/978-1-4666-9870-3.ch007

Srinivasan, S. (2016). Overview of Clinical Trial and Pharmacovigilance Process and Areas of Application of Computer System. In P. Chakraborty & A. Nagal (Eds.), *Software Innovations in Clinical Drug Development and Safety* (pp. 1–13). Hershey, PA: IGI Global. doi:10.4018/978-1-4666-8726-4.ch001

Srisawasdi, N. (2016). Motivating Inquiry-Based Learning Through a Combination of Physical and Virtual Computer-Based Laboratory Experiments in High School Science. In M. Urban & D. Falvo (Eds.), *Improving K-12 STEM Education Outcomes through Technological Integration* (pp. 108–134). Hershey, PA: IGI Global. doi:10.4018/978-1-4666-9616-7.ch006

Stavridi, S. V., & Hamada, D. R. (2016). Children and Youth Librarians: Competencies Required in Technology-Based Environment. In J. Yap, M. Perez, M. Ayson, & G. Entico (Eds.), *Special Library Administration, Standardization and Technological Integration* (pp. 25–50). Hershey, PA: IGI Global. doi:10.4018/978-1-4666-9542-9.ch002

Sung, W., Ahn, J., Kai, S. M., Choi, A., & Black, J. B. (2016). Incorporating Touch-Based Tablets into Classroom Activities: Fostering Children's Computational Thinking through iPad Integrated Instruction. In D. Mentor (Ed.), *Handbook of Research on Mobile Learning in Contemporary Classrooms* (pp. 378–406). Hershey, PA: IGI Global. doi:10.4018/978-1-5225-0251-7.ch019

Syväjärvi, A., Leinonen, J., Kivivirta, V., & Kesti, M. (2017). The Latitude of Information Management in Local Government: Views of Local Government Managers. *International Journal of Electronic Government Research*, *13*(1), 69–85. doi:10.4018/IJEGR.2017010105

Tanque, M., & Foxwell, H. J. (2018). Big Data and Cloud Computing: A Review of Supply Chain Capabilities and Challenges. In A. Prasad (Ed.), *Exploring the Convergence of Big Data and the Internet of Things* (pp. 1–28). Hershey, PA: IGI Global. doi:10.4018/978-1-5225-2947-7.ch001

Teixeira, A., Gomes, A., & Orvalho, J. G. (2017). Auditory Feedback in a Computer Game for Blind People. In T. Issa, P. Kommers, T. Issa, P. Isaías, & T. Issa (Eds.), *Smart Technology Applications in Business Environments* (pp. 134–158). Hershey, PA: IGI Global. doi:10.4018/978-1-5225-2492-2.ch007

Thompson, N., McGill, T., & Murray, D. (2018). Affect-Sensitive Computer Systems. In M. Khosrow-Pour, D.B.A. (Ed.), Encyclopedia of Information Science and Technology, Fourth Edition (pp. 4124-4135). Hershey, PA: IGI Global. doi:10.4018/978-1-5225-2255-3.ch357

Trad, A., & Kalpić, D. (2016). The E-Business Transformation Framework for E-Commerce Control and Monitoring Pattern. In I. Lee (Ed.), *Encyclopedia of E-Commerce Development, Implementation, and Management* (pp. 754–777). Hershey, PA: IGI Global. doi:10.4018/978-1-4666-9787-4.ch053

Triberti, S., Brivio, E., & Galimberti, C. (2018). On Social Presence: Theories, Methodologies, and Guidelines for the Innovative Contexts of Computer-Mediated Learning. In M. Marmon (Ed.), *Enhancing Social Presence in Online Learning Environments* (pp. 20–41). Hershey, PA: IGI Global. doi:10.4018/978-1-5225-3229-3.ch002

Tripathy, B. K. T. R., S., & Mohanty, R. K. (2018). Memetic Algorithms and Their Applications in Computer Science. In S. Dash, B. Tripathy, & A. Rahman (Eds.), Handbook of Research on Modeling, Analysis, and Application of Nature-Inspired Metaheuristic Algorithms (pp. 73-93). Hershey, PA: IGI Global. doi:10.4018/978-1-5225-2857-9.ch004

Turulja, L., & Bajgoric, N. (2017). Human Resource Management IT and Global Economy Perspective: Global Human Resource Information Systems. In M. Khosrow-Pour (Ed.), *Handbook of Research on Technology Adoption, Social Policy, and Global Integration* (pp. 377–394). Hershey, PA: IGI Global. doi:10.4018/978-1-5225-2668-1.ch018

Unwin, D. W., Sanzogni, L., & Sandhu, K. (2017). Developing and Measuring the Business Case for Health Information Technology. In K. Moahi, K. Bwalya, & P. Sebina (Eds.), *Health Information Systems and the Advancement of Medical Practice in Developing Countries* (pp. 262–290). Hershey, PA: IGI Global. doi:10.4018/978-1-5225-2262-1.ch015

Vadhanam, B. R. S., M., Sugumaran, V., V., V., & Ramalingam, V. V. (2017). Computer Vision Based Classification on Commercial Videos. In M. S., & V. V. (Eds.), Multi-Core Computer Vision and Image Processing for Intelligent Applications (pp. 105-135). Hershey, PA: IGI Global. doi:10.4018/978-1-5225-0889-2.ch004

Valverde, R., Torres, B., & Motaghi, H. (2018). A Quantum NeuroIS Data Analytics Architecture for the Usability Evaluation of Learning Management Systems. In S. Bhattacharyya (Ed.), *Quantum-Inspired Intelligent Systems for Multimedia Data Analysis* (pp. 277–299). Hershey, PA: IGI Global. doi:10.4018/978-1-5225-5219-2.ch009

Vassilis, E. (2018). Learning and Teaching Methodology: "1:1 Educational Computing. In K. Koutsopoulos, K. Doukas, & Y. Kotsanis (Eds.), *Handbook of Research on Educational Design and Cloud Computing in Modern Classroom Settings* (pp. 122–155). Hershey, PA: IGI Global. doi:10.4018/978-1-5225-3053-4.ch007

Wadhwani, A. K., Wadhwani, S., & Singh, T. (2016). Computer Aided Diagnosis System for Breast Cancer Detection. In Y. Morsi, A. Shukla, & C. Rathore (Eds.), *Optimizing Assistive Technologies for Aging Populations* (pp. 378–395). Hershey, PA: IGI Global. doi:10.4018/978-1-4666-9530-6.ch015

Wang, L., Wu, Y., & Hu, C. (2016). English Teachers' Practice and Perspectives on Using Educational Computer Games in EIL Context. *International Journal of Technology and Human Interaction, 12*(3), 33–46. doi:10.4018/IJTHI.2016070103

Watfa, M. K., Majeed, H., & Salahuddin, T. (2016). Computer Based E-Healthcare Clinical Systems: A Comprehensive Survey. *International Journal of Privacy and Health Information Management, 4*(1), 50–69. doi:10.4018/IJPHIM.2016010104

Weeger, A., & Haase, U. (2016). Taking up Three Challenges to Business-IT Alignment Research by the Use of Activity Theory. *International Journal of IT/Business Alignment and Governance, 7*(2), 1-21. doi:10.4018/IJITBAG.2016070101

Wexler, B. E. (2017). Computer-Presented and Physical Brain-Training Exercises for School Children: Improving Executive Functions and Learning. In B. Dubbels (Ed.), *Transforming Gaming and Computer Simulation Technologies across Industries* (pp. 206–224). Hershey, PA: IGI Global. doi:10.4018/978-1-5225-1817-4.ch012

Williams, D. M., Gani, M. O., Addo, I. D., Majumder, A. J., Tamma, C. P., Wang, M., ... Chu, C. (2016). Challenges in Developing Applications for Aging Populations. In Y. Morsi, A. Shukla, & C. Rathore (Eds.), *Optimizing Assistive Technologies for Aging Populations* (pp. 1–21). Hershey, PA: IGI Global. doi:10.4018/978-1-4666-9530-6.ch001

Wimble, M., Singh, H., & Phillips, B. (2018). Understanding Cross-Level Interactions of Firm-Level Information Technology and Industry Environment: A Multilevel Model of Business Value. *Information Resources Management Journal, 31*(1), 1–20. doi:10.4018/IRMJ.2018010101

Wimmer, H., Powell, L., Kilgus, L., & Force, C. (2017). Improving Course Assessment via Web-based Homework. *International Journal of Online Pedagogy and Course Design, 7*(2), 1–19. doi:10.4018/ IJOPCD.2017040101

Wong, Y. L., & Siu, K. W. (2018). Assessing Computer-Aided Design Skills. In M. Khosrow-Pour, D.B.A. (Ed.), Encyclopedia of Information Science and Technology, Fourth Edition (pp. 7382-7391). Hershey, PA: IGI Global. doi:10.4018/978-1-5225-2255-3.ch642

Wongsurawat, W., & Shrestha, V. (2018). Information Technology, Globalization, and Local Conditions: Implications for Entrepreneurs in Southeast Asia. In P. Ordóñez de Pablos (Ed.), *Management Strategies and Technology Fluidity in the Asian Business Sector* (pp. 163–176). Hershey, PA: IGI Global. doi:10.4018/978-1-5225-4056-4.ch010

Yang, Y., Zhu, X., Jin, C., & Li, J. J. (2018). Reforming Classroom Education Through a QQ Group: A Pilot Experiment at a Primary School in Shanghai. In H. Spires (Ed.), *Digital Transformation and Innovation in Chinese Education* (pp. 211–231). Hershey, PA: IGI Global. doi:10.4018/978-1-5225-2924-8.ch012

Yilmaz, R., Sezgin, A., Kurnaz, S., & Arslan, Y. Z. (2018). Object-Oriented Programming in Computer Science. In M. Khosrow-Pour, D.B.A. (Ed.), Encyclopedia of Information Science and Technology, Fourth Edition (pp. 7470-7480). Hershey, PA: IGI Global. doi:10.4018/978-1-5225-2255-3.ch650

Yu, L. (2018). From Teaching Software Engineering Locally and Globally to Devising an Internationalized Computer Science Curriculum. In S. Dikli, B. Etheridge, & R. Rawls (Eds.), *Curriculum Internationalization and the Future of Education* (pp. 293–320). Hershey, PA: IGI Global. doi:10.4018/978-1-5225-2791-6.ch016

Yuhua, F. (2018). Computer Information Library Clusters. In M. Khosrow-Pour, D.B.A. (Ed.), Encyclopedia of Information Science and Technology, Fourth Edition (pp. 4399-4403). Hershey, PA: IGI Global. doi:10.4018/978-1-5225-2255-3.ch382

Zare, M. A., Taghavi Fard, M. T., & Hanafizadeh, P. (2016). The Assessment of Outsourcing IT Services using DEA Technique: A Study of Application Outsourcing in Research Centers. *International Journal of Operations Research and Information Systems*, 7(1), 45–57. doi:10.4018/IJORIS.2016010104

Zhao, J., Wang, Q., Guo, J., Gao, L., & Yang, F. (2016). An Overview on Passive Image Forensics Technology for Automatic Computer Forgery. *International Journal of Digital Crime and Forensics*, 8(4), 14–25. doi:10.4018/IJDCF.2016100102

Zimeras, S. (2016). Computer Virus Models and Analysis in M-Health IT Systems: Computer Virus Models. In A. Moumtzoglou (Ed.), *M-Health Innovations for Patient-Centered Care* (pp. 284–297). Hershey, PA: IGI Global. doi:10.4018/978-1-4666-9861-1.ch014

Zlatanovska, K. (2016). Hacking and Hacktivism as an Information Communication System Threat. In M. Hadji-Janev & M. Bogdanoski (Eds.), *Handbook of Research on Civil Society and National Security in the Era of Cyber Warfare* (pp. 68–101). Hershey, PA: IGI Global. doi:10.4018/978-1-4666-8793-6.ch004

About the Contributors

Bhavya Alankar, Ph.D. (CSE), M.Tech. (CSE), Fellow (IETE), is currently working as a senior faculty at Department of Computer Science and Engg. at Jamia Hamdard, New Delhi, India. Previously employed at National Institute of Technology (NIT), Jalandhar, India. He has done his Ph.D. in Reconfigurable Computing from Uttarakhand Technical University, India and Masters in Technology in VLSI design from CDAC, Mohali, India and. He has 15 years of teaching and research experience. His research interests are in VLSI design, Cloud computing, Deep learning, Reconfigurable computing. He is author and editor to books in the area on VLSI, Machine Learning and IoT. He has received many awards and recognition from International bodies.

Harleen Kaur who hails from India, works in a Hamdard University, Department of Computer Science, New Delhi. She has served in United Nations University - IIGH as a Research Fellow focusing on Information Technology in Healthcare in 2011-2012. She holds a Ph.D. in Computer Science on the topic of Applications and Social Impact of Data Mining techniques in Health care Management. She has previously taught in University of Delhi, New Delhi, India. She researches broadly in the fields of Information Technology in healthcare, Knowledge Discovery as well as Data Mining and its applications. She has published numerous research articles in refereed international journals and conference proceedings and books. She is a member of several international bodies like ACUNS. She is an Editor to Springer book on "ICTs and the Millennium Development Goals: A United Nations Perspective". She is currently involved in the research project on Building a Standard Nationwide Healthcare Knowledge Driven System, for better utilization of research in ICT and development of policy making in knowledge-driven system to improve the quality of life for many citizens in the developing countries.

Ritu Chauhan pursued her Ph.D. in Computer Science from Jamia Hamdard University, New Delhi, India on the topic of Applications of Data Mining techniques in Spatial Databases. She graduated from the University of Delhi, New Delhi. She has previously served as a Lecturer in Computer Science, IMI. Currently, she is an Assistant Professor at Amity Institute of Biotechnology, Amity University. She has published numerous research articles in refereed international journals and conference proceedings and chapters in an edited book. She is a member of several international bodies. Her main research interests are in the fields of Data analysis with applications to medical databases.

* * *

Sandhya Avasthi is an Assistant Professor in computer science and Engineering Department at Dr Abdul Kalam Technical University. She is currently pursuing PhD in information technology from AIIT, Amity University Noida. Her research interests include natural language processing, information extraction, information retrieval, data science and business intelligence. She has considerable teaching experience and is author of many published research papers in machine learning techniques.

Rathishchandra Gatti is currently a researcher in the field of energy harvesting for cyber-physical systems, wave energy and also working as a professor in the department of mechanical engineering at Sahyadri College of Engineering and Management, India. Rathishchandra R Gatti received B.E from NITK Surathkal and Ph.D from Curtin University, Perth, Australia in 2002 and 2014 respectively. His management degrees include MBA from IGNOU Delhi (2007), Post graduate diploma in Global Strategic Management from ICFAI (2008), Six Sigma Greenbelt (DMAIC and DFSS) from General Electric, USA and Graduate Certificate in Research Commercialization from Curtin University, Australia. After his graduation, he has worked in the field of automotive R&D and design at Arvin Meritor, SVI and Rapsri. He also served as a NPI program manager at General Electric Industrial in R&D programs related to North American power distribution products from 2004 to 2008.From 2009 to 2014 pursued his PhD in the area of electromagnetic energy harvesting at Curtin University, Australia. He later taught engineering design as a sessional academic from 2014 to 2015 in at Curtin. He currently has 1US Patent, 1 Canadian patent, 1 Mexican patent and 13 peer-reviewed publications. He is a recognized reviewer of Journal of Mechanical Systems and Signal Processing, IEEE Transactions on Magnetics and Walailak Journal of Science and Technology. Linkedin Profile:https://www.linkedin.com/in/rathish-chandra-gatti-bb20a86.

Sindhu Hak Gupta received Ph.D. degree from Faculty of Engineering and Technology, Uttarakhand Technical University, Dehradun, India. She is currently an Associate Professor with the Department of Electronics and Communication Engineering, Amity School of Engineering and Technology, Amity University, Noida, India. Her research works include Energy Efficient Wireless Sensor Networks, Reliable Body Area Networks and Cooperative Communication.

Titus Haiduwa is currently a Lecturer in the Department of Information Technology, School of Computing, University of Namibia. His Research Interest: ICT4D,HCI, Software Development and Information Security.

Valerianus Hashiyana is currently a Senior Lecturer at School of Computing and Head of Department: Computer Science, University of Namibia. His area of research are Cybersecurity, Networking, IOT, e-health, Next generation computing.

Gloria Iyawa is currently a Senior Lecturer at Faculty of Computing and Informatics, Namibia University of Science and Technology. Her area of research are Healthcare Information Systems, IoT, Information Systems, Information Security to mention a few.

Rachna Jain did her B.E (Hons.) from C.R. State College of Engineering, Murthal. She completed her M.Tech in Information Technology from Guru Gobind Singh Indraprastha University, Delhi. She has completed her Ph.D in Adhoc Networks from Manav Rachna International Institute of Research and Studies, Faridabad. She is currently working as an Assistant Professor in the Department of Computer

Science & Engineering, JSS Academy of Technical Education, Noida. She has more than 19 years of teaching and research experience. Her research area includes mobile Adhoc networks, Security in wireless domain. Dr. Rachna Jain has been a state topper in Matriculation examination in Haryana Board in the year 1995. She has been topper throughout her academic journey earning many accolades and scholarships. She has also been a state level player of table tennis.

Suresh Joseph K. received his Bachelor Degree in Computer Science & Engineering from Bharathiar University and a Master Degree in Computer Science & Engineering from the University of Madras during 1999 and 2003 respectively. He received his Doctoral Degree in Computer Science & Engineering from Anna University in 2013. Presently he is working as an Associate Professor in the Department of Computer Science, Pondicherry University. His areas of research interest include Services Computing and Image Processing. He is having more than 50 research publications in reputed and Peer-reviewed journals.

Aditya Sam Koshy is currently pursuing Masters in Technology in the field of Computer Science from Jamia Hamdard University, New Delhi. He gained his Bachelors in Technology from the University of Kerala. Being a scholar of Computer Science and a rather curious person, he has always been keen towards the aforementioned discipline and other similitude fields, which attributed to him co-authoring the work "Security and Privacy Issues in Smart Cities- A Review".

Palanivel Kuppusamy has experience in the field of Software Engineering and Software Models & Architecture including research and development. He holds an M.Tech. Degree in which he has worked in the area of software architecture for e-Learning applications. He has published/presented more than 10 Book chapters and 40+ research papers and articles in international/national journals and conferences. He has an interest in writing articles related to data analytics, machine learning, and software models.

Anton Limbo is currently Lecturer at School of Computing under Information Technology Department, University of Namibia.. His areas of research are Internet of Things applications towards ICT for Development and Automation, Cyber security and High Performance Computing.

Syed Manzar did his Masters from Amity University, U.P., India.

Set-Sakeus Ndakolute is a student at School of Computing under Information Technology Department, University of Namibia. His area of interests are Internet of Things applications towards ICT for Development and Automation.

Maria Ntinda is currently Lecturer at School of Computing under Computer Science Department, University of Namibia.. Her areas of research are HCI, Programming, Educational Technologies, ICT for Indegenous Technology.

Nalina Suresh is currently a Senior Lecturer in the School of Computing, Department of Information technology University of Namibia. Her area of research are Networking and Security, Computational Theory and modelling, IOT, AI, Robotics, ML, DSP, educational technologies.

Martin Mabeifam Ujakpa is currently a Senior Lecturer & Faculty Dean at the International University of Management, Namibia. Previously he was a lecturer & Head of Academics with the Ghana Technology University College, Lecturer & Executive Manager at the Accra Institute of Technology, Ghana and Research Assistant & E-learning Administrator with the Valley View University, Ghana.

Index

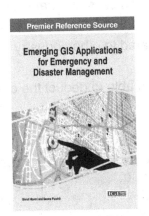

www.igi-global.com

Publisher of Peer-Reviewed, Timely, and Innovative Academic Research Since 1988

IGI Global's Transformative Open Access (OA) Model:
How to Turn Your University Library's Database Acquisitions Into a Source of OA Funding

Well in advance of Plan S, IGI Global unveiled their OA Fee Waiver (Read & Publish) Initiative. Under this initiative, librarians who invest in IGI Global's InfoSci-Books and/or InfoSci-Journals databases will be able to subsidize their patrons' OA article processing charges (APCs) when their work is submitted and accepted (after the peer review process) into an IGI Global journal.

How Does it Work?

Step 1: **Library Invests in the InfoSci-Databases:** A library perpetually purchases or subscribes to the InfoSci-Books, InfoSci-Journals, or discipline/subject databases.

Step 2: **IGI Global Matches the Library Investment with OA Subsidies Fund:** IGI Global provides a fund to go towards subsidizing the OA APCs for the library's patrons.

Step 3: **Patron of the Library is Accepted into IGI Global Journal (After Peer Review):** When a patron's paper is accepted into an IGI Global journal, they option to have their paper published under a traditional publishing model or as OA.

Step 4: **IGI Global Will Deduct APC Cost from OA Subsidies Fund:** If the author decides to publish under OA, the OA APC fee will be deducted from the OA subsidies fund.

Step 5: **Author's Work Becomes Freely Available:** The patron's work will be freely available under CC BY copyright license, enabling them to share it freely with the academic community.

Note: This fund will be offered on an annual basis and will renew as the subscription is renewed for each year thereafter. IGI Global will manage the fund and award the APC waivers unless the librarian has a preference as to how the funds should be managed.

Hear From the Experts on This Initiative:

"I'm very happy to have been able to make one of my recent research contributions *freely available* along with having access to the *valuable resources* found within IGI Global's InfoSci-Journals database."

— **Prof. Stuart Palmer,** Deakin University, Australia

"Receiving the support from IGI Global's OA Fee Waiver Initiative *encourages me to continue my research work without any hesitation.*"

— **Prof. Wenlong Liu,** College of Economics and Management at Nanjing University of Aeronautics & Astronautics, China

For More Information, Scan the QR Code or Contact:
IGI Global's Digital Resources Team at eresources@igi-global.com.

Are You Ready to Publish Your Research?

IGI Global
PUBLISHER of TIMELY KNOWLEDGE

IGI Global offers book authorship and editorship opportunities across 11 subject areas, including business, computer science, education, science and engineering, social sciences, and more!

Benefits of Publishing with IGI Global:

- Free one-on-one editorial and promotional support.
- Expedited publishing timelines that can take your book from start to finish in less than one (1) year.
- Choose from a variety of formats, including: Edited and Authored References, Handbooks of Research, Encyclopedias, and Research Insights.
- Utilize IGI Global's eEditorial Discovery® submission system in support of conducting the submission and double-blind peer review process.
- IGI Global maintains a strict adherence to ethical practices due in part to our full membership with the Committee on Publication Ethics (COPE).
- Indexing potential in prestigious indices such as Scopus®, Web of Science™, PsycINFO®, and ERIC – Education Resources Information Center.
- Ability to connect your ORCID iD to your IGI Global publications.
- Earn honorariums and royalties on your full book publications as well as complimentary copies and exclusive discounts.

Join Your Colleagues from Prestigious Institutions, Including:

Australian National University

MIT — Massachusetts Institute of Technology

Johns Hopkins University

Harvard University

Tsinghua University

Columbia University in the City of New York

Learn More at: www.igi-global.com/publish

or Contact IGI Global's Aquisitions Team at: acquisition@igi-global.com

Printed in the United States
by Baker & Taylor Publisher Services